AF360825

Distributed Systems and Mobile Computing

Distributed Systems and Mobile Computing

Editor

Giovanni Viglietta

MDPI • Basel • Beijing • Wuhan • Barcelona • Belgrade • Manchester • Tokyo • Cluj • Tianjin

Editor
Giovanni Viglietta
Japan Advanced Institute of
Science and Technology (JAIST)
Japan

Editorial Office
MDPI
St. Alban-Anlage 66
4052 Basel, Switzerland

This is a reprint of articles from the Special Issue published online in the open access journal *Information* (ISSN 2078-2489) (available at: https://www.mdpi.com/journal/information/special_issues/distributed_systems_mobile_computing).

For citation purposes, cite each article independently as indicated on the article page online and as indicated below:

LastName, A.A.; LastName, B.B.; LastName, C.C. Article Title. *Journal Name* **Year**, *Volume Number*, Page Range.

ISBN 978-3-0365-2842-7 (Hbk)
ISBN 978-3-0365-2843-4 (PDF)

Contents

About the Editor

Giovanni Viglietta has been an Assistant Professor in the School of Information Science at the Japan Advanced Institute of Science and Technology (JAIST) since 2018. He completed his PhD in Computer Science at the University of Pisa (Italy) in 2013, and he obtained postdoctoral fellowships from the University of Ottawa (Canada) and ETH Zurich (Switzerland) in 2013–2018. His research is at the interface between Mathematics and Theoretical Computer Science; his interests include Distributed Computing, Discrete and Computational Geometry, and Combinatorial Game Theory. Throughout his career, he has made contributions to several research areas, including motion planning for mobile robots, population protocols, programmable matter, computer vision, polyhedral combinatorics, the mathematical analysis of games and puzzles, etc. Dr. Viglietta has collaborated with several research teams from USA, Canada, Japan, France, The Netherlands, Italy, Mexico, etc., and has published two book chapters and over 50 peer-reviewed articles in international journals and conferences. In 2015, he was a recipient of the Best Paper Award at the 17th International Symposium on Stabilization, Safety, and Security of Distributed Systems (SSS '15) for a paper on population protocols. He has been a member of the IEICE since 2021; he has been in the Program or Organizing Committees of 11 international conferences and workshops, and has been a referee or sub-referee for over 40 international journals and conferences.

Preface to "Distributed Systems and Mobile Computing"

Recent years have seen rapid development of the field of Distributed Computing, whose interest is the study of systems of autonomous computational entities with limited communication capabilities. Since its first appearances in the 1960s and 1970s, in the context of concurrent processes in early operating systems, Distributed Computing has become a rich and diverse branch of theoretical computer science. Current applications are no longer limited to message-passing in concurrent processes, but extend to wireless computer and sensor networks, the Internet of Things (IoT), autonomous intelligent vehicles, swarms of mobile robots, programmable matter, crawlers on the Web, molecular interactions, etc. Research in Distributed Computing is now at the interface between graph theory, computational geometry, combinatorics, and control theory.

This book is aimed at specialists in computer science, mathematics, and engineering who are interested in current research developments in two topical areas of Distributed Computing: network construction and analysis, and motion planning for mobile robots. The volume includes five research papers authored by some established world-class experts in modern Distributed Computing.

The first article is about the recently introduced model of network constructors: groups of anonymous agents that interact randomly and have the ability to form connections. This behavior naturally emerges in situations where the environment exhibits passive or adversarial dynamics, while agents can actively modify the network structure; examples are found in molecular interactions, which occur in dynamic solutions where proteins are nonetheless able to form stable bonds. The article investigates the fundamental question of which families of networks can be stably constructed in polylogarithmic parallel time, assuming the presence of a unique leader in the system.

The second article is concerned with improving quality of service in IoT networks. While long-range communication offers several advantages in terms of flexibility and cost, ultra-dense IoT networks employing these technologies experience a significant increase in traffic intensity. This is mitigated by subdividing the network into clusters and selecting a structure that maximizes quality of service, as measured under some mathematical models. Typical metrics are throughput and channel capacity. This article formulates a mathematical model to estimate the expected throughput capacity of a given clustering choice. It also develops clustering methods, which depend on the distribution of end devices, achieving near-optimal throughput capacity.

The subject of the last three articles are mobile robots, which are autonomous entities capable of sensing and moving through the environment. Typically, such robots are modeled as points moving within a continuous geometric space, such as a line or a surface. The first of the three articles studies the classic gathering problem, in which all robots in a large swarm, initially scattered across a plane, are tasked with meeting at a common location. Agreeing on a gathering point is an important part of the problem; in fact, the robots do not have the ability to send messages to each other, and they do not have an explicit way to communicate. Another complicating factor is that robots have limited visibility, and only see other robots within a certain range. The article proposes a time-optimal gathering algorithm, assuming that all robots have compasses pointing in the same direction.

The two final articles deal with the evacuation problem, where two robots must locate an unknown exit point within a given region, and reach it in the shortest possible time. The first article studies the evacuation problem on a line, introducing a novel mechanic: the two robots share a single device, called "bike", that can be used by only one robot at a time to move at greater speed. Two models of communication are considered: face-to-face, in which the robots can communicate only when they physically meet, and wireless, in which they can always communicate instantaneously and at any distance. The article proposes several evacuation algorithms for both communication models, comparing their efficiency, and also discusses lower bounds.

The last article studies the evacuation problem on a disk in the face-to-face communication model, which is the most common setting for this problem. This article introduces a multi-objective version of the classical problem, where the robots must evacuate the disk within a specified time while minimizing the average evacuation time. The latter requirement is equivalent to designing a randomized algorithm with an optimal expected evacuation time. In addition to providing several algorithms that achieve different tradeoffs between the two objective functions, the authors introduce a new systematic method for performing both worst-case and average-case analysis for any evacuation algorithm that admits an analytic description. The method is then applied to previously known algorithms, as well as to the new algorithms proposed in this article.

I would like to thank MDPI, especially in the person of Managing Editor Amanda Liu, for giving me the opportunity to edit this book and for assisting me throughout the process.

Giovanni Viglietta
Editor

 *information*

MDPI

Article

On the Distributed Construction of Stable Networks in Polylogarithmic Parallel Time

Matthew Connor [1], Othon Michail [1] and Paul Spirakis [1,2,*]

[1] Department of Computer Science, University of Liverpool, Liverpool L69 3BX, UK; M.Connor3@liverpool.ac.uk (M.C.); Othon.Michail@liverpool.ac.uk (O.M.)
[2] Computer Engineering and Informatics Department, University of Patras, 265 04 Patras, Greece
[*] Correspondence: P.Spirakis@liverpool.ac.uk

Abstract: We study the class of networks, which can be created in polylogarithmic parallel time by *network constructors*: groups of anonymous agents that interact randomly under a uniform random scheduler with the ability to form connections between each other. Starting from an empty network, the goal is to construct a stable network that belongs to a given family. We prove that the class of trees where each node has any $k \geq 2$ children can be constructed in $O(\log n)$ parallel time with high probability. We show that constructing networks that are k-regular is $\Omega(n)$ time, but a minimal relaxation to (l, k)-regular networks, where $l = k - 1$, can be constructed in polylogarithmic parallel time for any fixed k, where $k > 2$. We further demonstrate that when the finite-state assumption is relaxed and k is allowed to grow with n, then $k = \log \log n$ acts as a threshold above which network construction is, again, polynomial time. We use this to provide a partial characterisation of the class of polylogarithmic time network constructors.

Keywords: population protocol; distributed network construction; polylogarithmic time protocol; spanning tree; regular network; partial characterisation

Citation: Connor, M.; Michail, O.; Spirakis, P. On the Distributed Construction of Stable Networks in Polylogarithmic Parallel Time. *Information* **2021**, *12*, 254. https://doi.org/10.3390/info12060254

Academic Editor: Giovanni Viglietta

Received: 1 May 2021
Accepted: 15 June 2021
Published: 19 June 2021

1. Introduction

Passively dynamic networks are an important type of dynamic network in which the network dynamics are *external* to the algorithm and are a property of the environment in which a given system operates. Wireless sensor networks in which individual sensors are carried by autonomous entities, such as animals, or are deployed in a dynamic environment, such as the flow of a river, are examples of passively dynamic networks. In terms of modelling such systems, the network dynamics are usually assumed to be controlled by an *adversary scheduler* who has exclusive control over the interaction or communication sequence among the computational entities.

One line of research assumes the scheduler to be *fair* in the sense that it can forever conceal potentially reachable configurations of the system. This sub-type of passively dynamic networks are known as population protocols and were introduced in the seminal paper of Angluin et al. [1]. A type of fair scheduler, which is typically assumed when the running time of protocols is to be analysed, is the *uniform random scheduler*, which in every discrete step, selects equiprobably a pair of entities to interact from all permissible pairs of entities. Traditionally, the population protocols literature considers extremely weak entities and the goal is to reveal the computational possibilities and limitations under such a challenging interaction scheme. Recent progress has highlighted the interesting trade-offs between local space of the entities and the running time of protocols, showing, among other things that very fast running times (where fast is, here, considered to be anything growing as polylog(n), n being the total number of entities in the system) can be achieved for a wide range of basic distributed tasks if the entities are equipped with as few states as polylog(n). Alistarh and Gelashvili [2] also proposed the first sub-linear leader election protocol, which stabilises in $O(\log^3 n)$ parallel time, assuming $O(\log^3 n)$ states at each agent. Gasieniec and Stachowiak [3] designed a space optimal ($O(\log\log n)$ states) leader election protocol, which stabilises in $O(\log^2 n)$ parallel time. General characterisations, including upper and lower bounds of the trade-offs between time and space in population protocols, are provided in [4]. Doty et al. [5] showed that a state count of $O(n^{60})$ enables fast and exact population counting.

Another line considers worst-case adversary schedulers, which may even be aware of the protocol and try to optimise against it. There, the entities are typically assumed to be powerful, such as processors of traditional distributed systems, and the only restrictions imposed on the scheduler are instantaneous or temporal connectivity restrictions, which essentially do not allow the scheduler to forever block communication between any two parts of the system. This was initiated by O'Dell and Wattenhofer [6] for the asynchronous case, and then the synchronous case was extensively studied in a series of papers by Kuhn et al. [7]. Michail et al. [8] extended this to the case of possibly disconnected dynamic networks in which connectivity is only guaranteed in a temporal sense.

The other main type of dynamic networks, with respect to who controls the changes in the network topology, are *actively dynamic networks*. In such networks, the algorithm is able to either implicitly change the sequence of interactions by controlling the mobility of the entities or explicitly modify the network structure by creating and destroying communication links at will. This is, for example, the subject of the area of overlay network construction [9–12]; very recently, Michail et al. introduced a fully distributed model for computation and reconfiguration in actively dynamic networks [13].

An interesting alternative family of dynamic networks rises when one considers a mixture of the passive network dynamics of the environment and the active dynamics resulting from an algorithm that can partially control the network changes or that can fix network structures that the environment is unable to affect. This is naturally motivated by molecular interactions where, for example, proteins can bind to each other, forming structures and maintaining their stability despite the dynamicity of the solution in which they reside. Michail and Spirakis [14] introduced and studied such an abstract model of distributed network construction, called the *network constructors* model, where the network dynamicity is the same as in population protocols but now the finite-state entities can additionally activate and deactivate pairwise connections upon their interactions. It was shown that very complex global networks can be formed stably, despite the dynamicity of the environment. Then, Michail [15] studied a geometric variant of network constructors in which the entities can only form geometrically constrained shapes in 2D or 3D space. Another interesting hybrid dynamic network model is the one by Gmyr et al. [16] in which the entities have partial control over the connections of an otherwise worst-case passively dynamic network, following the model of Kuhn et al. [7].

Our Approach

We investigate which families of networks can be stably constructed by a distributed computing system in polylogarithmic parallel time. To our knowledge, this is the first attempt made to approach this task.

Our protocols assume the existence of a leader node. A node x is a *leader node* if in the initial configuration, all $u \in V \setminus \{x\}$, where V is the set of all nodes, are in state q_0 and x is in state $s \neq q_0$.

We first study the *k-children spanning tree* problem, where the goal is to construct a tree where each node has, at most $k \geq 2$, children. We show that it is possible to solve this problem for any k in $O(\log n)$ time with high probability. We then show that network constructors which create k-regular graphs necessarily take $\Omega(n)$ time. However, with minimal relaxation to $(k, k-1)$-regular networks, the problem can be solved for any constant $k \geq 2$ in polylogarithmic time. We examine this as a special case of the (l, k)-*Regular Network* problem, where the goal is to construct a spanning network in which every node has at least $l < k$ and at most k connections, where $2 < k < n$. We then transition to experimental analysis of the protocol, which not only provides evidence of the sharp contrast of the minimal relaxation but also reveals a threshold value for k, beyond which the problem reverts to polynomial time. We use this knowledge to propose a first partial characterisation of the set of polylogarithmic time network constructors. We leave providing formal bounds as an open problem.

In Section 2, we formally define the model of network constructors and the network construction problems that are considered in this work. In Section 3, we study the k-children spanning tree problem and provide the lower bound for k-regular networks. We then present a protocol for the (l, k)-regular network problem and our experimental analysis, culminating in partial characterisation. In Section 4, we conclude and give further research directions that are opened by our work.

2. Materials and Methods

2.1. The Model

Definition 1. *A Network Constructor (NET) is a distributed protocol defined by a 4-tuple $(Q, q_0, Q_{out}, \delta)$, where Q is a finite set of node-states, $q_0 \in Q$ is the initial node-state, $Q_{out} \subseteq Q$ is the set of output node-states, and $\delta : Q \times Q \times \{0, 1\} \rightarrow Q \times Q \times \{0, 1\}$ is the transition function.*

If $\delta(a, b, c) = (a', b', c')$, we call $(a, b, c) \rightarrow (a', b', c')$ a *transition* (or *rule*) and we define $\delta_1(a, b, c) = a', \delta_2(a, b, c) = b'$, and $\delta_3(a, b, c) = c'$. A transition $(a, b, c) \rightarrow (a', b', c')$ is called *effective* if $x \neq x'$ for at least one $x \in \{a, b, c\}$ and *ineffective* otherwise. When we present the transition function of a protocol, we only present the effective transitions. Additionally, we agree that the *size* of a protocol is the number of its states, i.e., $|Q|$.

The system consists of a population V_I of n distributed *processes* (called *nodes* for the rest of this paper). In the generic case, there is an underlying *interaction graph* $G_I = (V_I, E_I)$ specifying the permissible interactions between the nodes. Interactions in this model are always pairwise. In this work, G_I is a *complete undirected interaction graph*, i.e., $E_I = \{uv : u, v \in V_I \text{ and } u \neq v\}$, where $uv = \{u, v\}$. Initially, all nodes in V_I are in the initial node-state q_0. A central assumption of the model is that edges have binary states. An edge in state 0 is said to be *inactive* while an edge in state 1 is said to be *active*. All edges are initially inactive. Execution of the protocol proceeds in discrete steps. In every step, a pair of nodes uv from E_I is selected by an *adversary scheduler* and these nodes interact and update their states and the state of the edge joining them, according to the transition function δ.

A *configuration* is a mapping $C : V_I \cup E_I \rightarrow Q \cup \{0, 1\}$ specifying the state of each node and each edge of the interaction graph. Let C and C' be configurations, and let u, v be distinct nodes. We say that C goes to C' via *encounter* $e = uv$, denoted $C \xrightarrow{e} C'$, if $(C'(u), C'(v), C'(e)) = \delta(C(u), C(v), C(e))$ or $(C'(v), C'(u), C'(e)) = \delta(C(v), C(u), C(e))$ and $C'(z) = C(z)$, for all $z \in (V_I \setminus \{u, v\}) \cup (E_I \setminus \{e\})$. We say that C' *is reachable in one step from C*, denoted $C \rightarrow C'$, if $C \xrightarrow{e} C'$ for some encounter $e \in E_I$. We say that C' is *reachable* from C and write $C \rightsquigarrow C'$, if there is a sequence of configurations $C = C_0, C_1, \ldots, C_t = C'$, such that $C_i \rightarrow C_{i+1}$ for all $i, 0 \leq i < t$.

An *execution* is a finite or infinite sequence of configurations $C_0, C_1, C_2, \ldots$, where C_0 is an initial configuration and $C_i \rightarrow C_{i+1}$, for all $i \geq 0$. A *fairness condition* is imposed on the adversary to ensure that the protocol makes progress. An infinite execution is *fair* if for every pair of configurations C and C' such that $C \rightarrow C'$, if C occurs infinitely often in the execution then so does C'. In what follows, every execution of a NET is, by definition, considered to be fair.

We define the *output of a configuration* C as the graph $G(C) = (V, E)$ where $V = \{u \in V_I : C(u) \in Q_{out}\}$ and $E = \{uv : u, v \in V, u \neq v, \text{ and } C(uv) = 1\}$. In words, the output graph of a configuration consists of those nodes that are in output states and those edges between them that are active, i.e., the active subgraph induced by the nodes that are in output states. The output of an execution $C_0, C_1, \ldots$ is said to *stabilize* (or *converge*) to a graph G if there exists some step $t \geq 0$ such that (abbreviated "s.t." in several places) $G(C_i) = G$ for all $i \geq t$, i.e., from step t and onwards, the output graph remains unchanged. Every such configuration C_i, for $i \geq t$ is called *output stable*. The *running time* (or *time to convergence*) of an execution is defined as the minimum, such as t (or ∞ if no such t exists). Throughout the paper, whenever we study the running time of a NET, we assume that interactions are chosen by a *uniform random scheduler*, which, in every step, selects independently and uniformly at random one of the $|E_I| = n(n-1)/2$ possible interactions. In this case, the running time becomes a random variable (abbreviated "r.v." throughout) X and our goal is to obtain bounds on the expectation $E[X]$ of X. Note that the uniform random scheduler is fair with probability 1.

In this work, "time" is treated as sequential in our analyses, i.e., a time step consists of a single interaction selected by the scheduler. Such a sequential estimate can be easily translated to some estimate of parallel time. For example, assuming that $\Theta(n)$ interactions occur in parallel in every step, one could obtain an estimation of parallel time by dividing sequential time by n. All results are given in parallel time.

Definition 2. *We say that an execution of a NET on n nodes constructs a graph (or network) G, if its output stabilises to a graph isomorphic to G.*

Definition 3. *We say that a protocol P constructs a graph language L, if in every execution P constructs a graph $G \in L$ and for all G, there exists an execution of P, which constructs G.*

2.2. Problem Definitions

Here, we provide formal definitions for all of the classes of networks considered in this paper.

k-Children Spanning Tree: The goal is to construct a spanning tree where each individual element has at most $k \in \mathbb{N}$ children.
(l,k)-Regular Network: A spanning network where for any $l, k \in \mathbb{N}$ where $l < k$, elements with degree $d < l$ form a clique and all others have a degree of at least l and at most k.

2.3. Experimental Setup

We performed experiments with the goal of guiding a proof of the running time necessary to solve the (l,k)-regular network problem. We learned that a formal proof would be difficult, due to the reliance of random variables on the values of other random variables, so we left this as an open problem. We then experimented with different values of k to see what the effect would be, and discovered a running time threshold in the process. All were implemented using C and compiled with GCC. All tests were repeated at least five times per the value of n and the average number of time steps were taken as the result. To terminate our experiments, we designed special stabilisation conditions.

3. Results

3.1. Polylogarithmic Time Protocols for k-Children Spanning Trees

In this section, we study the complexity of the k-Children Spanning Tree problem. We give a protocol (Algorithm 1) and show that it has a running time of $O(\log n)$ parallel time with high probability.

Algorithm 1 *k-Slot protocol*

$Q = \{F, L_0, L_1, \ldots, L_k, O_0, O_1, \ldots, O_k\}$
δ:

$(L_x, F, 0) \rightarrow (L_{x+1}, O_0, 1)$ for $x < k$
$(O_y, F, 0) \rightarrow (O_{y+1}, O_0, 1)$ for $y < k$

In the above protocol, the F state corresponds to being a node, which is not a member of the tree. L_i corresponds to the *leader* node, which acts as the root of the tree, and O_i to non-leader nodes in the tree, where i represents the number of children of a given node.

We assume that for every execution of Algorithm 1 on a population P of n nodes, $n - 1$ nodes initialise to the state F and one node initialises to the state L_0.

We begin by proving that Algorithm 1 stably constructs the graph language $T_k = \{G | G$ is a rooted tree and $\forall u \in P \implies \Delta^+(u) \leq k\}$, where $\Delta^+(u)$ is defined as the number of children of the node u and $k \geq 2$. We then show that this is accomplished in $O(\log n)$ parallel time. Then, we show that this bound holds with high probability.

Lemma 1. *Under Algorithm 1, the connected component S, defined as the leader node and all nodes connected to the leader either directly or indirectly through some other nodes, is eventually spanning.*

Proof. We observe that the number of open slots o is initially k. o is non-decreasing, as every increase in $\Delta^+(u)$ for some u necessarily increases $|V(S)|$. Since there are always open slots available, every unconnected node is guaranteed to be able to connect to S at some point. Therefore, when S stabilises, it contains all $u \in P$. $\square$

Lemma 2. *For all executions of Algorithm 1 on the population P of n nodes, it stabilises to some $G \in T_k$ where $|V(G)| = n$.*

Proof. We prove this via an induction on the connected component S. For the base case, there is one node in the state L_0. This is trivially a member of T_k, as no connections have formed yet. We now assume that there is a connected component of size $|S|$. For a connected component of size $|S| + 1$, an unconnected node $u \in V \setminus S$ in the state F must connect to S at some node $x \in S$. By Lemma 1, such a node must exist. If the node x has two children, it is in the state O_2 or L_2, as for all nodes in states O_i and L_j, the i and j correspond to the number of children of those nodes. Since there are no defined transitions from these states, no u can connect to x. Therefore, S remains a tree and $G(S) \in T_k$. $\square$

Lemma 3. *For all $G \in T_k$, there is an execution of Algorithm 1, which stabilises on G when starting on a population P of size $n = |V(G)|$.*

Proof. We first set the value of k to the maximum number of connections in any node in the tree. Let the leader node l in the population P correspond to the root r of G. If r has i children, connect i nodes in the state F to l. For each child c of the leader node, let it correspond to a child d of r. If d has j children, connect j nodes in the state F to c. Continuing this process for all nodes $u \in G$, the result is a spanning tree where all nodes in the tree are equivalent to some $u \in G$. $\square$

Theorem 1. *Algorithm 1 stably constructs the graph language T_k in $O(\log n)$ time w.h.p.*

Proof. By application of the Lemmas above. $\square$

We now show that Algorithm 1 constructs T_k in $O(\log n)$ time w.h.p by considering executions where $k = 2$. Executions where $k > 2$ are necessarily faster, as they have more open slots per node.

Lemma 4. *Let $G \in T_2$ of n nodes. The number of available nodes $\alpha(G) = \lfloor |G|/2 \rfloor + 1$.*

Proof. Observe that for G, every second node that connects to G keeps the number of available nodes the same. This is because two new nodes must become children of the same node, and the second new node takes the second slot. For the base case, $n = 1$ and $\alpha = 0 + 1 = 1$. We divide $n = n + 1$ into two cases: n is *even* and n is *odd*. If n is *even*, then $\alpha = n/2 + 1$. Then, for $n = n + 1$, $\alpha = \lfloor n + 1/2 \rfloor + 1 = n/2 + 1$. This corresponds to the observation earlier that every other node (i.e., n is *odd*) should not increase α. If n is *odd*, then $\alpha = \lfloor n/2 \rfloor + 1 = (n-1)/2 + 1$. Then, for $n = n + 1$, $\alpha = \lfloor n + 1/2 \rfloor + 1 = n/2 + 1$ as expected. $\square$

Remark 1. *At any point during the execution of Algorithm 1, for the connected component S,* $G(S) \in \alpha(G)$.

Let the probablistic process P be an execution of Algorithm 1 for $k = 2$ with the following scheduling restriction: If at any point during the execution of Algorithm 1 two nodes x and y have exactly one child, disconnect that child of x or y, which is a leaf, and connect it to the other node. If both are leaves, pick one at random.

Lemma 5. *The expected time to convergence of the probabalistic process P is $O(\log n)$.*

Proof. Let the r.v. X be the number of steps until convergence. A step is *successful* if any unconnected node joins the connected component S. An *epoch i* is the period beginning with the step following the $(i-1)$st success and ending with the step at which the ith success occurs. The r.v. X_i, $1 \leq i \leq n-1$ is the number of steps in epoch i. p_i is the probability of success at any step in epoch i. This is defined as $p_i = \frac{2\alpha(T_i)(n-i)}{n(n-1)}$, where T_i is the graph of the strongly connected component $G(S)$ in epoch i.

It follows that $E[X_i] = 1/p_i = \frac{n(n-1)}{2\alpha(T_i)(n-i)}$. By linearity of expectation we have the following:

$$E[X] = E\left[\sum_{i=1}^{n-1} X_i\right] = \sum_{i=1}^{n-1} E[X_i] = \sum_{i=1}^{n-1} \frac{n(n-1)}{2\alpha(T_i)(n-i)} = \frac{n(n-1)}{2} \sum_{i=1}^{n-1} \frac{1}{\alpha(T_i)(n-i)}$$

$$= \frac{n(n-1)}{2} \sum_{i=1}^{n-1} \frac{1}{(\lfloor i/2 \rfloor + 1)(n-i)} \leq \frac{n(n-1)}{2} \sum_{i=1}^{n-1} \frac{1}{(i/2)(n-i)}$$

$$= n(n-1) \sum_{i=1}^{n-1} \frac{1}{i(n-i)} = n(n-1) \sum_{i=1}^{n-1} \frac{1}{n}\left(\frac{1}{i} + \frac{1}{n-i}\right)$$

$$= (n-1)\left[\sum_{i=1}^{n-1} \frac{1}{i} + \sum_{i=1}^{n-1} \frac{1}{n-i}\right] = (n-1)2H_{n-1} = 2(n-1)[\ln(n-1) + O(1)]$$

$$= O(n \log n)$$

$\square$

Lemma 6. *The running time of P is the worst case running time for Algorithm 1.*

Proof. Assume that there is an execution of A, which has an slower running time than P. Such an execution must have a lower number of available nodes at some point than P. If the execution simulates the scheduling restriction of P, then it cannot be slower than P. If the execution does not simulate the restriction, then at some point, two nodes have two leaves and one is not shifted to the other. The number of available nodes is, therefore, greater by one and the expected running time is faster than P. Therefore, any execution of A must be at least as fast as P. $\square$

Theorem 2. *The expected running time of Algorithm 1 is upper bounded by the $O(\log n)$ running time of P.*

Proof. By application of Lemmas 5 and 6. $\square$

Lemma 7. *For each time step in Algorithm 1, the probability of any node in the set of unconnected nodes U connecting to the tree is at least $2\frac{|U|}{n}$.*

Proof. Assume that there are $|S|$ nodes, which are connected to the tree. The probability of a node $x \in U$ connecting to the tree is $\frac{|S||U|}{n^2}$. If there are at least $n/2$ nodes connected to the tree, then $\frac{|S||U|}{n^2} \geq \frac{1/2|U|}{n^2} = 2\frac{|U|}{n}$. The case where there are less than $n/2$ nodes connected to the tree is symmetrical, meaning that same process happens in reverse for $1 \leq n \leq n/2$. Therefore, $2\frac{|U|}{n}$ is a lower bound of the probability of connecting to the tree. $\square$

Lemma 8. *For Algorithm 1, the number of time steps until convergence is $O(\log n)$ w.h.p.*

Proof. Consider the scenario where m balls are being thrown into n bins. If Z is the random variable for the number of empty bins, then $E[Z] = n(1 - 1/n)^m$. The probability of a ball entering an empty bin is e/n; e is the number of empty bins. For our protocol scenario, balls are time steps and bins are unconnected nodes. So $m = an\ln n$ is the number of balls, and n is the number of bins. Since the probabilty of success in the balls and bins scenario is lower than $2\frac{|U|}{n}$ when $|U| = e$, we can use it as a bound for the probability of success. Therefore, $E[Z] = n(1 - 1/n)^{an\ln n} \leq ne^{-an\ln n} = n^{1-a}$. Using Markov's inequality, $E[Z \geq 1] \leq E[Z] = \frac{1}{n^a}$. Since a can be set arbitrarly high, convergence is $O(\log n)$ w.h.p. $\square$

Theorem 3. *Algorithm 1 stably constructs the graph language T_k in $O(\log n)$ time w.h.p.*

Proof. By application of Lemmas 7 and 8. $\square$

3.2. Time Thresholds for (l, k)-Regular Networks

In this section, we present our solution for the (l, k)-Regular Network problem for $l = k - 1$, the *Cross-edges Tree* protocol. We first show that a *k-regular network*, defined as a network where each node has a degree exactly equal to k, cannot be constructed in polylogarithmic time. We then show via experimental analysis that this impossibility result does not hold for the minimal relaxation of (l, k)-Regular Networks when k is a constant and $l = k - 1$. Finally, we demonstrate that when k exceeds the threshold of $\log \log n$, the protocol itself is no longer in the polylogarithmic time class. Note that from now on, k refers to the *degree* of a node, not the *number of children*.

Theorem 4. *Any protocol which constructs a k-regular network where $k < n$ has a running time of $\Omega(n)$.*

Proof. Consider the population P of size n, using a generic k-regular network construction protocol X. The number of connections is limited by k to $\frac{kn}{2}$, as this is less than the $\frac{n(n-1)}{2}$ maximum for n nodes. The population initially has kn network connection entry points, which can be used to make new connections and which decrease by 2 for every connection made. Since $(kn) \leq n(n-1)$, at some point in the execution, there must be two nodes with 1 unused entry point each. Using these points and stabilising the protocol means that both nodes must be selected by the scheduler at the same time, an event with probability $\frac{1}{n^2}$. Since an event with probability $\frac{1}{n^2}$ is unavoidable, the protocol X must construct a network in at least $\Omega(n^2)$ interactions. $\square$

In light of the above impossiblity, we now give our protocol (Algorithm 2) for the (l, k)-Regular Network problem when $l = k - 1$. Note that the leader node has at most two connections; this is to guarantee that there is no scenario where a partial network with unconnected nodes cannot connect to forms.

Algorithm 2 *Cross-edges Tree*

$Q = \{F, L_0, L_1, \ldots, L_k, O_0, O_1, \ldots, O_k\}$
$\delta:$

$(L_x, F, 0) \rightarrow (L_{x+1}, O_0, 1)$ for $x < k$
$(O_y, F, 0) \rightarrow (O_{y+1}, O_0, 1)$ for $y < k$
$(L_x, O_y, 0) \rightarrow (L_{x+1}, O_{y+1}, 1)$ for $x, y < (k-1)$
$(O_y, O_z, 0) \rightarrow (O_{y+1}, O_{z+1}, 1)$ for $y, z < (k-1)$

The Cross-edges Tree protocol adds additional rules, allowing leaves within a tree to connect to other nodes within the tree as though they are candidates for becoming children.

Stabilisation Conditions of the (l, k)-Regular Network

To implement a simulator that can provide results efficiently, we had to define and prove conditions which, when fulfilled, ensure that the protocol is stable.

Lemma 9. *For $n > 3$, protocol 2 stabilises with at least 1 node, which is not in the state of k in the connected tree.*

Proof. If all nodes in the connected tree are in the O_k state, then at some point, two O_{k-1} nodes would have to change to the O_k state, which is against the rules of the protocol. $\square$

Corollary 1. *Due to the presence of nodes in a state $s \neq O_k$ and the fairness condition, Algorithm 2 never stabilises with isolated nodes, defined as nodes which are not part of the main tree structure.*

Lemma 10. *For $n > 3$, protocol 2 stabilises with, at most, $k - 2$ nodes in the states $\{O_x | x < k - 1\}$.*

Proof. Assume that there are $k - 1$ nodes in the states O_x. If this is the case, there must be some node that is connected to the tree and every other node with any state O_x; otherwise, the protocol is not stable. This node must have the state O_{k-1}, which is not in O_x. Therefore, the number of nodes in O_x is, at most, $k - 2$. $\square$

Theorem 5. *For $n > k > 3$, at most, $n - k - 2$ nodes have a degree of either k or $k - 1$ and $l \leq k - 2$ nodes are of a degree of at least 1 and at most $k - 2$.*

Proof. All nodes with degree $x < k - 1$ must be members of a clique; otherwise, the protocol is not stable. By Lemma 12, we know that there are no isolated nodes, or nodes with degree $d < 1$. By Lemma 13, we know that there are, at most, $k - 2$ nodes of degree less than $k - 1$, and that the maximum state within the clique is O_{k-2}. Therefore, the theorem must hold. $\square$

We now provide the results of simulating the protocol for $k = 3$. We used the same conditions as in the other running time experiments, executing the protocol 10 times for each population size n, where $n = 10 + 6t$, where t is the test number from 0 to 199. The results are given in Figure 1.

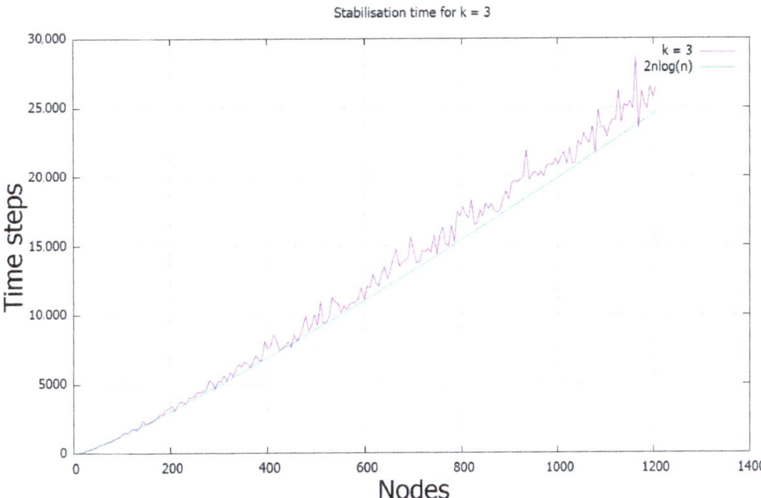

Figure 1. Running time of the protocol for $k = 3$, compared with a polylogarithmic function.

The running time is difficult to prove formally. This is because random variables are used, which represent the number of nodes with a given degree in a given time step. Their values depend on the values of all random variables in the previous time step. We, therefore, turn our focus to experiments based on measuring the impact of the value of k on the running time of the protocol.

We measured the running time of our Cross-edges Tree protocol for different network sizes. The results below (Figure 2) show that a higher value of k has little effect on the running time until k exceeds $\log \log n$.

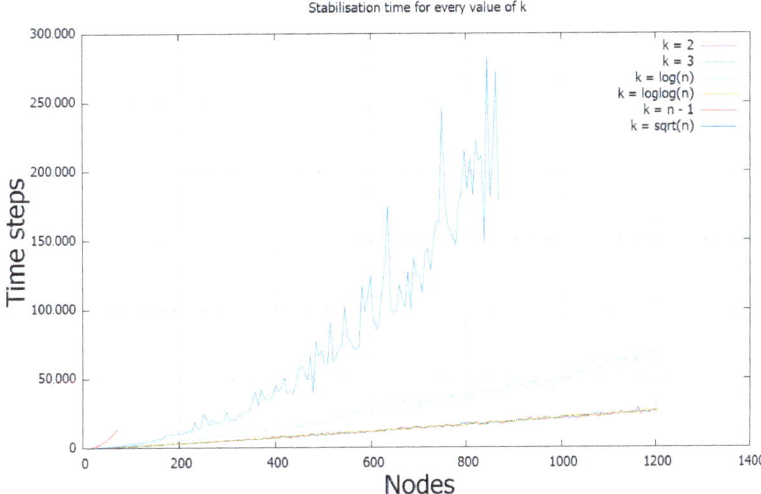

Figure 2. The effect of k on the running time of the protocol.

To investigate why the protocol slows down dramatically after this point, we ran experiments where we stored the number of nodes with specific degrees in each time step. We executed the protocol with 200 nodes, and ran 10 iterations. These degrees were set to $0, 1, k/2, k-1,$ and k. We collected results for $k = \log \log n$ (Figure 3), $k = \log n$ (Figure 4), and $k = \sqrt{n}$ (Figure 5). The results show that the cause seems to be a large reduction in the number of nodes, which are in the $k - 1$ state as k grows as a fraction of n. They suggest that when the fraction of $k - 1$ nodes is below some fraction between $1/4$ and $1/8$ of the total, the protocol slows down and enters the class of protocols with polynomial time.

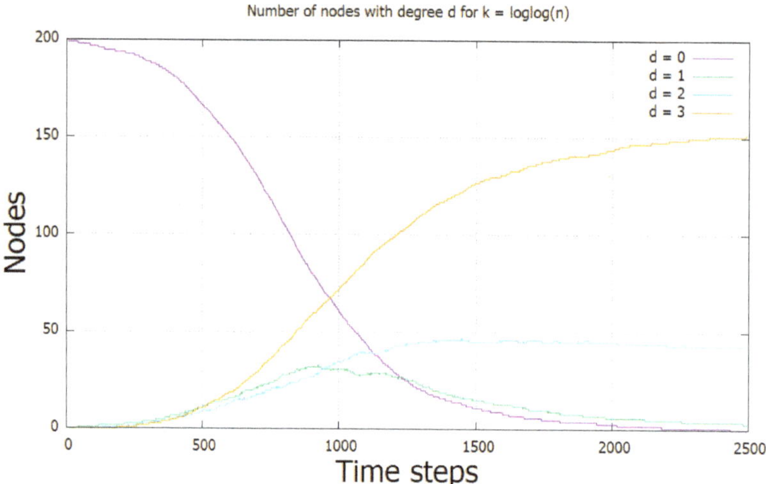

Figure 3. The results for $k = \log \log n$. Note the difference in the axes labels.

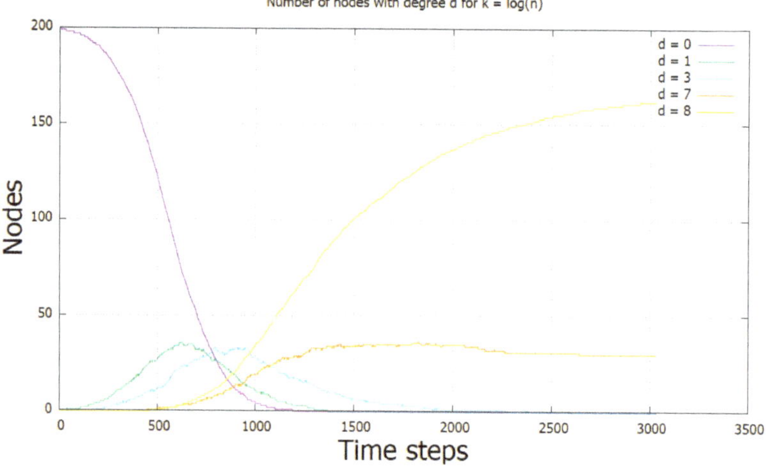

Figure 4. The results for $k = \log n$. Here, we see the beginning of a leftwards shift of the lines, and an upwards shift in $d = k$.

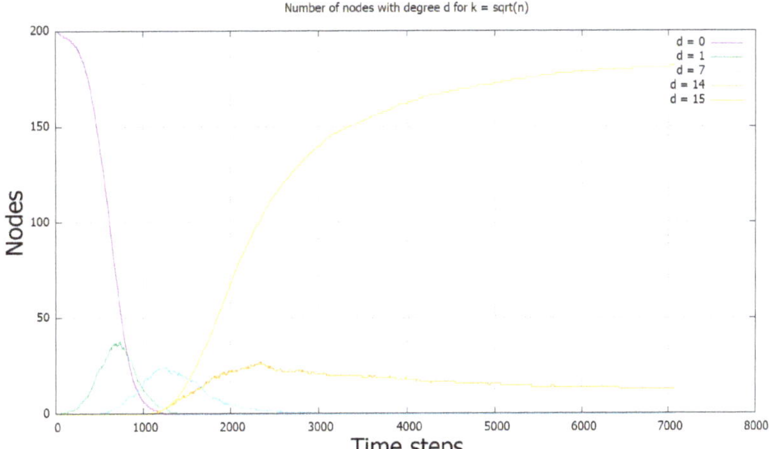

Figure 5. The results for $k = \sqrt{n}$. Both shifts are more intense.

4. Conclusions

Population protocols have a history of being applied in the context of physical devices [17]. However, to our knowledge, there has yet to be an attempt to apply the network constructors model to such a setting. For a cluster of devices, as in an Internet of Things style network [18], it may be desirable to form networks in order to solve problems in a very time-sensitive and highly dynamic context. This should typically be achieved without each device being aware of the size of the network or what specific devices the others are connected to. Network constructors and, thus, the protocols presented in this work, may be may be applicable to this, as they help define the class of networks which could be constructed in such a context. Works on social networks and social communities are a possible application domain for our models (see, e.g., [19]). It would be interesting to further investigate the potential applicability of population protocols and network constructors in data streaming applications; a preliminary attempt to do this was attempted by [20].

There are a number of open problems to be addressed. The most important is to develop an exact characterisation of the class of networks which can be constructed in polylogarthimic parallel time. However, there are other, more immediate problems. For example, we have yet to investigate the effect that widening the difference between k and l will have on the protocol. We speculate that this will result in a faster running time in exchange for less uniformity within the resulting spanning network. We also speculated about the possibilities of using a leaderless version of our Cross tree Protocol. We believe that such a protocol may offer a trade off between the running time and the possibility of forming networks, which are spanning, depending on the values of k and l.

Author Contributions: Conceptualization, O.M.; Data curation, M.C.; Formal analysis, M.C. and O.M.; Investigation, M.C.; Methodology, O.M.; Software, M.C.; Supervision, O.M.; Writing—original draft, M.C.; Writing—review and editing, M.C., O.M. and P.S. All authors have read and agreed to the published version of the manuscript.

Funding: This research was funded by an EPSRC Doctoral Training Studentship.

Data Availability Statement: We did not use any real data from data sources.

Conflicts of Interest: The authors declare no conflict of interest.

References

1. Angluin, D.; Aspnes, J.; Diamadi, Z.; Fischer, M.J.; Peralta, R. Computation in networks of passively mobile finite-state sensors. *Distrib. Comput.* **2006**, *18*, 235–253. [CrossRef]
2. Dan Alistarh, R.G. Polylogarithmic-time leader election in population protocols. In Proceedings of the 42nd International Colloquium on Automata, Languages, and Programming (ICALP), Kyoto, Japan, 6–10 July 2015; pp. 479–491.
3. Gasieniec, L.; Stachowiak, G. Fast Space Optimal Leader Election in Population Protocols. In Proceedings of the 2018 Annual ACM-SIAM Symposium on Discrete Algorithms (SODA18), New Orleans, LA, USA, 7–10 January 2018; pp. 2653–2667.
4. Alistarh, D.; Aspnes, J.; Eisenstat, D.; Gelashvili, R.; Rivest, R.L. Time-space trade-offs in population protocols. In Proceedings of the 2017 Annual ACM-SIAM Symposium on Discrete Algorithms (SODA17), Barcelona, Spain, 16–19 January 2017; pp. 2560–2579.
5. Doty, D.; Eftekhari, M.; Michail, O.; Spirakis, P.G.; Theofilatos, M. Brief Announcement: Exact Size Counting in Uniform Population Protocols in Nearly Logarithmic Time. In Proceedings of the 32nd International Symposium on Distributed Computing (DISC), Schloss Dagstuhl–Leibniz-Zentrum fuer Informatik, New Orleans, LA, USA, 15–19 October 2018; pp. 46:1–46:3. [CrossRef]
6. O'Dell, R.; Wattenhofer, R. Information dissemination in highly dynamic graphs. In Proceedings of the 2005 Joint Workshop on Foundations of Mobile Computing (DIALM-POMC), Cologne, Germany, 2 September 2005; pp. 104–110. [CrossRef]
7. Kuhn, F.; Lynch, N.; Oshman, R. Distributed computation in dynamic networks. In Proceedings of the Forty-Second ACM Symposium on Theory of Computing (STOC), Cambridge, MA, USA, 5–8 June 2010; pp. 513–522.
8. Michail, O.; Chatzigiannakis, I.; Spirakis, P.G. Causality, Influence, and Computation in Possibly Disconnected Synchronous Dynamic Networks. In Proceedings of the 16th International Conference on Principles of Distributed Systems (OPODIS), Rome, Italy, 18–20 December 2012; pp. 269–283.
9. Angluin, D.; Aspnes, J.; Chen, J.; Wu, Y.; Yin, Y. Fast Construction of Overlay Networks. In Proceedings of the 17th ACM symposium on Parallelism in Algorithms and Architectures (SPAA), Las Vegas, NV, USA, 18–20 July 2005; pp. 145–154.
10. Aspnes, J.; Shah, G. Skip Graphs. *ACM Trans. Algorithms (TALG)* **2007**, *3*, 37. [CrossRef]
11. Aspnes, J.; Wu, Y. $O(\log n)$-Time Overlay Network Construction from Graphs with Out-Degree 1. In Proceedings of the 11th International Conference on Principles of Distributed Systems (OPODIS), Guadeloupe, France, 17–20 December 2007; pp. 286–300.
12. Götte, T.; Hinnenthal, K.; Scheideler, C. Faster Construction of Overlay Networks. In Proceedings of the 26th International Colloquium on Structural Information and Communication Complexity (SIROCCO), L'Aquila, Italy, 1–4 July 2019; pp. 262–276.
13. Michail, O.; Skretas, G.; Spirakis, P.G. Distributed Computation and Reconfiguration in Actively Dynamic Networks. In Proceedings of the 39th ACM Symposium on Principles of Distributed Computing (PODC), Virtual Event, Italy, 3–7 August 2020; pp. 448–457.
14. Michail, O.; Spirakis, P.G. Simple and efficient local codes for distributed stable network construction. *Distrib. Comput.* **2016**, *29*, 207–237. [CrossRef]
15. Michail, O. Terminating distributed construction of shapes and patterns in a fair solution of automata. *Distrib. Comput.* **2018**, *31*, 343–365. [CrossRef]
16. Gmyr, R.; Hinnenthal, K.; Scheideler, C.; Sohler, C. Distributed Monitoring of Network Properties: The Power of Hybrid Networks. In Proceedings of the 44th International Colloquium on Automata, Languages, and Programming (ICALP), Warsaw, Poland, 10–14 July 2017; pp. 137:1–137:15.
17. Becchetti, L.; Bergamini, L.; Ficarola, F.; Salvatore, F.; Vitaletti, A. First Experiences with the Implementation and Evaluation of Population Protocols on Physical Devices. In Proceedings of the 2012 IEEE International Conference on Green Computing and Communications, Besancon, France, 20–23 November 2012; pp. 335–342. [CrossRef]
18. Atzori, L.; Lera, A.; Morabito, G. The Internet of Things: A survey. *Comput. Netw.* **2010**, *54*, 2787–2805. [CrossRef]
19. Holub, S.; Khymytsia, N.; Holub, M.; Fedushko, S. The Intelligent Monitoring of Messages on Social Networks. *CEUR Workshop Proc.* **2020**, *2616*, 308–317.
20. Àlvarez, C.; Chatzigiannakis, I.; Duch, A.; Gabarró, J.; Michail, O.; Serna, M.; Spirakis, P.G. Computational models for networks of tiny artifacts: A survey. *Comput. Sci. Rev.* **2011**, *5*, 7–25. [CrossRef]

 *information*

Article

Clustering Optimization of LoRa Networks for Perturbed Ultra-Dense IoT Networks

Mohammed Saleh Ali Muthanna [1,2], Ping Wang [3], Min Wei [3,*], Ahsan Rafiq [1] and Nteziriza Nkerabahizi Josbert [1]

[1] School of Computer Science and Technology, Chongqing University of Posts and Telecommunications, Chongqing 400065, China; muthanna@mail.ru (M.S.A.M.); L201710003@stu.cqupt.edu.cn (A.R.); L201710007@stu.cqupt.edu.cn (N.N.J.)

[2] Department of Automation and Control Processes, Saint Petersburg Electrotechnical University "LETI", 197022 Saint Petersburg, Russia

[3] School of Automation, Chongqing University of Posts and Telecommunications, Chongqing 400065, China; wangping@cqupt.edu.cn

* Correspondence: weimin@cqupt.edu.cn

Abstract: Long Range (LoRa) communication is widely adapted in long-range Internet of Things (IoT) applications. LoRa is one of the powerful technologies of Low Power Wide Area Networking (LPWAN) standards designed for IoT applications. Enormous IoT applications lead to massive traffic results, which affect the entire network's operation by decreasing the quality of service (QoS) and minimizing the throughput and capacity of the LoRa network. To this end, this paper proposes a novel cluster throughput model of the throughput distribution function in a cluster to estimate the expected value of the throughput capacity. This paper develops two main clustering algorithms using dense LoRa-based IoT networks that allow clustering of end devices according to the criterion of maximum served traffic. The algorithms are built based on two-common methods, K-means and FOREL. In contrast to existing methods, the developed method provides the maximum value of served traffic in a cluster. Results reveal that our proposed cluster throughput model obtained a higher average throughput value by using a normal distribution than a uniform distribution.

Keywords: Internet of Things; dense networks; LPWAN; LoRa; clustering; throughput; capacity; QoS

Citation: Muthanna, M.S.A.; Wang, P.; Wei, M.; Rafiq, A.; Josbert, N.N. Clustering Optimization of LoRa Networks for Perturbed Ultra-Dense IoT Networks. *Information* **2021**, *12*, 76. https://doi.org/10.3390/info12020076

Academic Editor: Giovanni Viglietta

Received: 25 January 2021
Accepted: 8 February 2021
Published: 10 February 2021

Publisher's Note: MDPI stays neutral with regard to jurisdictional claims in published maps and institutional affiliations.

1. Introduction

Currently, in the era of the IoT [1], communication technologies have significantly expanded to reach a variety of industries. Consequently, providing low-power and long-distance communication networks has become an essential component of many applications within smart cities, such as waste management [2], supply chain [3], Industrial Internet of Things [4], smart metering [5], and traffic control. Low Power Wide Area Networking (LPWAN) is an effective way for such applications to overcome these cost, energy, and complexity challenges, especially when such applications require covering large geographic areas. One of the attracted extensive attentions for LPWAN technologies today is Long Range (LoRa) [6]. The communication model of LoRa can be considered a better alternative for short-range and cellular communications in different applications and will offer notable features, such as low data rates, long-range, and low power consumption. The LoRa technology provides a high flexibility network by introducing long-range communication at low cost and communication specifications (e.g., bit rate, throughput, and delay). Network flexibility is an important issue that is related to LoRa network design [7].

However, a dense number of deployed devices in a large geographical area lead to a significant increase in subscriber traffic intensity. The demand from large retailers is growing to offer enhanced quality of services (QoS) and channel capacity, which is difficult for network operators [8]. The main trends in the development of LoRa networks consist of the rise of LoRa capacity and the number of users served.

13

In some cases, these processes lead to a decrease in the QoS requirements. To ensure the required QoS in LoRa networks' design, it is necessary to resort to various methods to guarantee the sufficiency of the relevant resources, such as reducing the service area of LoRa gateways, increasing their number, and maximizing the throughput and capacity of the LoRa networks [9]. It can be shown that the previous problems are highly related and simultaneously affect QoS in LoRa networks performance. These problems should be solved considering the peculiarities of the placement of end devices in the service area, possibly considering their movements. When organizing LoRa end devices connection, it is required to solve the problem of ensuring QoS traffic within the cluster and between network elements and certain elements of clusters.

The clustering methods consist of selecting a certain number of clusters and selecting a structure that provides the maximum possible traffic QoS. The clustering problem's solution is similar to the optimization problem's solution in which a certain metric is minimized (maximized). Such a metric may be throughput and channel capacity.

This paper analyzes the channel capacity between the cluster member (CM) and the cluster head (CH). We develop a cluster throughput model to estimate the expected value of the throughput capacity and develop clustering methods to make a rational choice of the algorithm depending on the distribution of end devices, which allows obtaining a cluster throughput capacity value close to the maximum. The main contributions of this paper are summarized as follows:

- The channel capacity analysis between the CM and the CH showed its dependence on the distribution of end devices. Remarkably, the results have shown that a larger average throughput is achieved with a normal distribution than with a uniform distribution;
- A cluster throughput model has been developed to estimate the throughput capacity's expected value when forming the cluster of end devices, which allows using it in the end devices' clustering problems;
- Clustering methods have been developed to make a rational choice of the algorithm depending on the distribution of end nodes, which allows obtaining a cluster throughput capacity value close to the maximum.

The rest of the paper is presented in several sections: Section 2 presents the related work and motivation. The LoRa technology overview is detailed in Section 3. Section 4 illustrates the problem statement. Section 5 presents the system description. Section 6 presents the clustering methods. Evaluation results based on two-common methods, K-means and FOREL, are presented in Section 7. Section 8 offers the discussion, and lastly, in Section 9, we make a conclusion and discuss future work.

2. Related Work and Motivation

Recent works on LoRa and Long Rang Wide Area Networking (LoRaWAN) have mainly dealt with LoRa performance evaluation in terms of capability [10,11], performance [12], lifetime [13], latency [14], and parameter setting [15,16] for industrial monitoring applications. For example, in [7], a temperature monitoring application is proposed. Dynamic Line Rating (DLR) for an Overhead Transmission Line (OTL) system is monitored as it relies on the weather, temperature, and inclination measurements. Here, OTL monitoring is carried by a vision system, which is further transmitted by LoRa communication. Communication is carried between the vision system and supervisory control and data acquisition (SCADA) system. This paper illustrates that QoS has a vital function in the received data reliability.

The paper in [17] aims to determine the scalability of the LoRa technology. For that, a high density of sensors is deployed in IoT environments for smart city applications. When the network scale is increased, the data transmission system has been affected. This paper finalizes that the data transmission system can be improved by focusing on optimal transmission policies, including the spreading factor (SF) allocation. Significantly, SF allocation is affected by distance with gateway and other factors. With an increase in network scale, a single gateway fails to ensure better data transmission efficiency.

The work in [18] constructs an infrastructure that is capable of tracking and monitoring the environmental system. LoRa technology is used for tracking applications to overcome energy-related limitations. As this system is planned to be implemented in coastal applications, multiple gateways are deployed. With the multiple gateways, this work attains the required level of connectivity and coverage in the network. Received signal strength indicator (RSSI) is considered for SF allocation, and the data is encrypted for security reasons.

The paper in [19] proposed an adaptive data rate (ADR) algorithm for improving error performance in rough channels resulting in extending the range of the network, the problem addressed by this paper to enhance the scalability, robustness, and fairness between nodes of the network by reducing the number of the data messages in the up-link direction as well as the medium access control (MAC) command messages in the down-link direction by using ADR algorithm. ADR algorithm in this paper is effective in stable channel conditions with a small-scale network, but the main motive of LoRa is in highly variable conditions. Thus, ADR algorithm in this work does not include in large complex networks.

To improve QoS in LoRa, the authors in [20] first derived the IoT node's mathematical model. From the mathematical model, the closed-form formula is derived to characterize the node performance. Then, the performance is maximized by optimizing the performance of IoT nodes. For formulating nodes performance, the Markov chain model is utilized. Then, the optimal transmission policy is derived based on the mathematical model of the node. The work in [20] considered both types of data, including normal data and emergency data. For both data, performance is optimized by a genetic algorithm (GA) and simulated annealing (SA) algorithm. Although normal and emergency types are considered, the same fitness function is formulated for both types. However, different types of data require different QoS levels, which is not achieved in this work. GA and SA update the parameters. GA is complex in nature and difficult to handle scalable problems, while SA is very slow and sensitive to even small changes in the input values. Thus, the optimal parameter assignment for all packets by GA and SA is relatively tricky. Parameters are updated by clustering algorithms, which are efficient since it learns the environment continuously.

Complementary to the presented related works, this paper focuses particularly on throughput capacity in LoRa networks and provides a solution to maximize throughput capacity value and build solutions to achieve the QoS requirements by developing clustering algorithms for distributing and managing LoRa gateways for smart city application and IoT dense networks.

3. LoRa Technology Overview

In [21], the proprietary LoRa physical layer (PHY) technique is possessed via Semtech Corporation. In [22], LoRa Alliance has specified LoRaWAN as the medium access control (MAC) layer protocol. This section described more specific details of LoRa technology, LoRaWAN, and the main characteristics in the following subsection.

3.1. LoRa Physical Layer

The LoRa modulation has several parameters: (i) spreading factor (SF); (ii) bandwidth (BW); (iii) chirp spread spectrum (CSS); code rate (CR) [23]. In LoRa modulation, information is transmitted in symbols, the length of which T_s depends on using (SF). Each symbol is a sinusoidal signal, the frequency of which is cyclically shifted within a bandwidth (BW).

In LoRa modulation, the symbol duration T_s and bit rate (R_b) can be calculated as following [24]:

$$R_b = SF * \frac{BW}{2^{SF}} * CR, \tag{1}$$

$$T_s(s) = \frac{2^{SF}}{bw}, \tag{2}$$

where SF refers to the spreading factor, and BW is bandwidth. The transmitted symbol rate Rs is calculated as $Rs(Symbol/sec) = \frac{1}{Ts} = \frac{bw}{2^{SF}}$. Thus, the chip rate Rc can be defined as $Rc(chips/sec) = Rs * 2^{SF}$, since, as we previously stated, $Rc = bw$.

LoRa modulation also comprises a variable error correction scheme that enhances the signal transmission robustness at the expense of redundancy. Therefore, the data nominal bit rate, Rb, can be defined as the following [23]:

$$R_b(bps) = \frac{SF * bw}{2^{SF}}\left(\frac{4}{4 + CR}\right), \tag{3}$$

where CR is for error correction and equal to 4/5, 4/6, 4/7, and 4/8.

Another critical parameter is the receiver sensitivity, which indicates the lowest power level of the received LoRa signal that the receiver can detect and demodulate. Based on the LoRa Semtech designer's guide, the receiver sensitivity of LoRa can be calculated as $\rho(dBm) = -174 + 10logBW + NF + SNR$.

Where ρ is the receiver sensitivity, NF is the receiver's noise figure, and SNR is the signal-to-noise ratio of the received signal. Table 1 indicates the nominal bit rate and the receiver sensitivity for the bandwidth of 125 kHz. For the values in Table 1, the maximum communication range of the LoRa is around 10 km.

Table 1. Transmission speed and receiver sensitivity from spreading factor (SF).

Bandwidth (kHz)	SF	Nominal Bit Rate R_b (bps)	Sensitivity (ρ)(dBm)
125	6	9375	−118
125	7	5469	−123
125	8	3125	−126
125	9	1758	−129
125	10	977	−132
125	11	537	−134
125	12	293	−137

As a LoRa network operates at a frequency of 868 MHz, bandwidth of 125 kHz, the payload of 8 bytes, and a preamble of 6, the payload of 8 bytes, and a preamble of 6, the number of symbols in the physical layer data block can be specified as following [24]:

$$pqyloadSymNb = 8 + \max\left(\text{ceil}\left(\frac{8PL - 4SF + 28 + 16CRC - 20H}{4(SF - 2DE)}\right)(CR + 4), 0\right), \tag{4}$$

where ceil (x) maps to the smallest integer that is greater than the value of x, SF is the spreading factor, CRC is the cyclic redundancy check, H is the header mode, DE is the data rate, CR is the coding rate, and PL is the number of payload bytes of the physical layer block and can be determined based on the payload of the application layer FRM as

$$PL = 12 + FRM$$

where FRM is the payload of the application layer.

CRC denotes the payload's existence, and it is fixed to either 0 or 1 to refer to the off and on statuses. The header mode is also fixed to either 0 or 1; H = 0, when the explicit header mode is utilized, and H = 1 when the implicit header mode is authorized.

The total duration of the LoRa frame T_{frame} can be calculated of the sum of the transmission time of the preamble $T_{preamble}$ and the payload $T_{payload}$ as the following:

$$T_{frame} = T_{preamble} + T_{payload}. \tag{5}$$

The preamble time can be calculated as follows:

$$T_{preamble} = \left(n_{preamble} + 4,25\right) * T_s, \tag{6}$$

where $n_{preamble}$ is the programmable length of modem registers.

The payload duration can be calculated as follows:

$$T_{payload} = payloadSymNb * T_s. \tag{7}$$

Summing up, the total frame duration can be calculated by adding both Equations (6) and (7).

$$T_{frame} = \left(n_{preamble} + 4,25\right) * T_s + payloadSymNb * T_s. \tag{8}$$

Thus, the value of the frame duration changes with the different values of the spreading factor. Table 2 indicates the different values of the up-link frame duration time at the different values of the SF used by LoRa systems. Furthermore, Table 3 shows the same for one confirmatory frame.

Table 2. Values of up-link frame duration at various values of SF used by Long Range (LoRa).

SF	bw (kHz)	T_s (ms)	$n_{preamble}$	FRM (byte)	PL (byte)	H	CRC	DE	CR	Payload-SymNb	$T_{preamble}$ (ms)	$T_{payload}$ (ms)	TUL_{frame} (ms)
6	125	0.51	6	8	20	0	1	0	1	48	5.25	24.48	29.73
7	125	1.02	6	8	20	0	1	0	1	43	10.5	43.86	54.36
8	125	2.05	6	8	20	0	1	0	1	38	20.99	77.9	98.89
9	125	4.1	6	8	20	0	1	0	1	33	41.98	135.3	177.28
10	125	8.19	6	8	20	0	1	0	1	33	83.97	270.27	354.24
11	125	16.38	6	8	20	0	1	0	1	28	167.94	458.64	626.58
12	125	32.77	6	8	20	0	1	0	1	28	335.87	917.56	1253.43

Table 3. Calculation of the transmission time of one confirmatory frame.

SF	bw (kHz)	T_s (ms)	$n_{preamble}$	FRM (byte)	PL (byte)	H	CRC	DE	CR	Payload-SymNb	$T_{preamble}$ (ms)	$T_{pay-load}$ (ms)	TUL_{frame} (ms)
6	125	0.51	6	0	12	1	1	0	1	28	5.25	14.28	19.53
7	125	1.02	6	0	12	1	1	0	1	28	10.5	28.56	39.06
8	125	2.05	6	0	12	1	1	0	1	23	20.99	47.15	68.14
9	125	4.1	6	0	12	1	1	0	1	23	41.98	94.3	136.28
10	125	8.19	6	0	12	1	1	0	1	18	83.97	147.42	231.39
11	125	16.38	6	0	12	1	1	0	1	18	167.94	294.84	462.78
12	125	32.77	6	0	12	1	1	0	1	18	335.87	589.86	925.73

3.2. LoRaWAN MAC Layer

LoRaWAN is a network protocol designed for many LPWAN applications that use unlicensed frequency bands for transmission. Its standard was published in 2015 and describes the data link layer protocol, while the physical layer protocol is proprietary and belongs to the transmitter manufacturer. Figure 1 shows the defined protocol of LoRaWAN via the LoRa Alliance [22].

LoRaWAN devices are categorized into three classes: A, B, and C. Class A is primary, based on the "asynchronous ALOHA" [25] access method, and is required to be supported by all devices. All LoRaWAN sensors, when turned on, work according to class A and can switch to other classes if such a physical possibility is available and upon agreement with the server. Class B is based on the periodic distribution of service information from the server and access to the channel on a schedule. This assumes that devices can consume more power than Class A devices, which will allow them to listen to periodic messages from the server. Class C is based on constant listening of the channel by sensors.

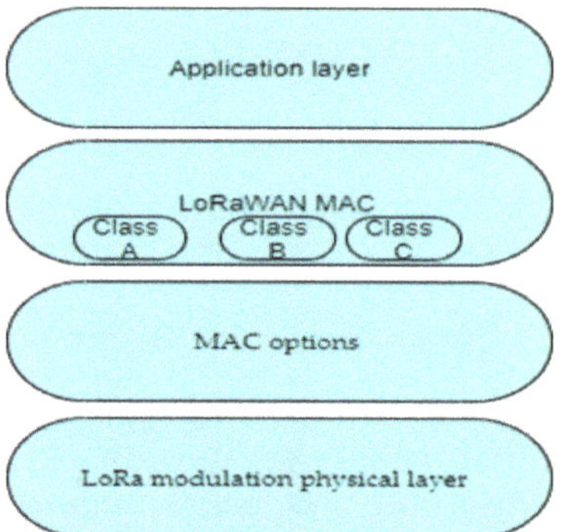

Figure 1. LoRaWAN (Long Range Wide Area Network) stack.

A LoRaWAN network architecture contains end devices, network servers, and LoRa gateways, as shown in Figure 2 [26]. The connection between the gateways and the server is reliable and fast, and the gateways are connected to sensors wirelessly using LoRa technology. The server is the coordinator of the network, and the gateways play the role of repeaters between the sensors and the server—having received a frame via a wireless connection, the gateway encapsulates the frame in an IP packet and transmits it to the server and, similarly, transmits the packets from the server to the sensors.

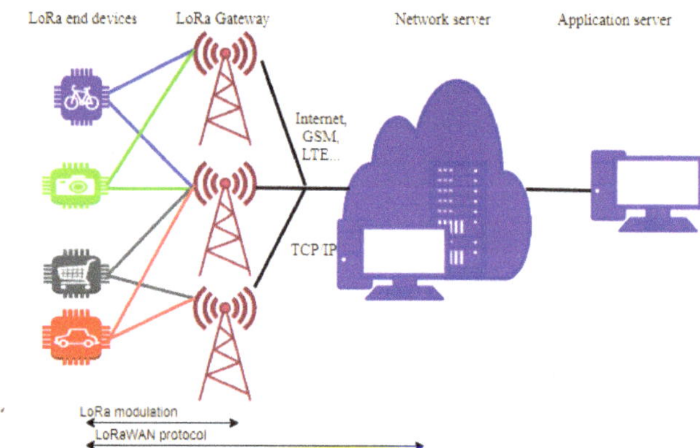

Figure 2. LoRaWAN network architecture.

3.3. Calculation of Packet Arrival Rate

Packets from different end nodes arrive at the gateways in a Poisson process [27]. In turn, the gateway receives the packets and, as a response, transmits a confirmation packet. Unconfirmed packets are re-transmitted, also forming a Poisson stream. Since a huge number of nodes are located in the network, thus the probability that several nodes transmit simultaneously is high. This probability can be calculated as follows:

The probability of the fact that during the transmission of one packet T in the air, there are still k packets from other nodes is determined as

$$P(k) = \frac{\lambda^k \exp^{-\lambda}}{k!},$$ (9)

where P (k) is the probability of k packets in the air, transmitted in parallel, and λ is the arrival rate of packets in time T. When the number of parallel packets on the air is zero, i.e., k = 0, there is no collision. The packet is successfully transmitted to the base station or the gateway. In this case, the probability of successful transmission of a packet, $P_{successful}$, is defined as

$$P_{successful} = \exp^{-2\lambda}. \tag{10}$$

Therefore, the probability of unsuccessful transmission, which is indicated as the loss probability, can be calculated as

$$P_{loss}(\lambda) = 1 - \exp^{-2\lambda}, \tag{11}$$

where P_{loss} is the loss probability that also indicates the collision probability. The effect of variations of the rate of arrival packets on the collision probability is illustrated in Figure 3.

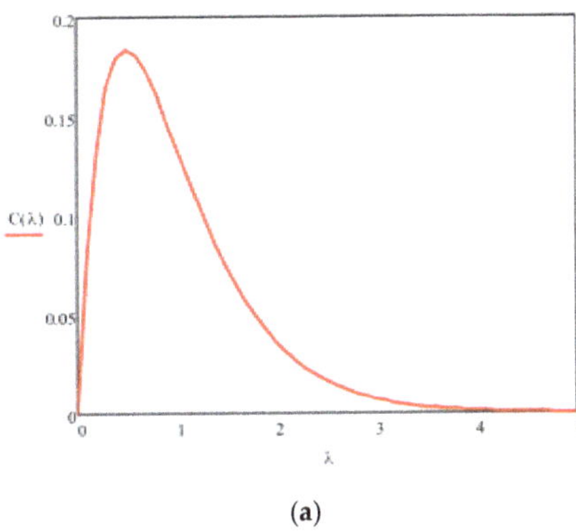

(a)

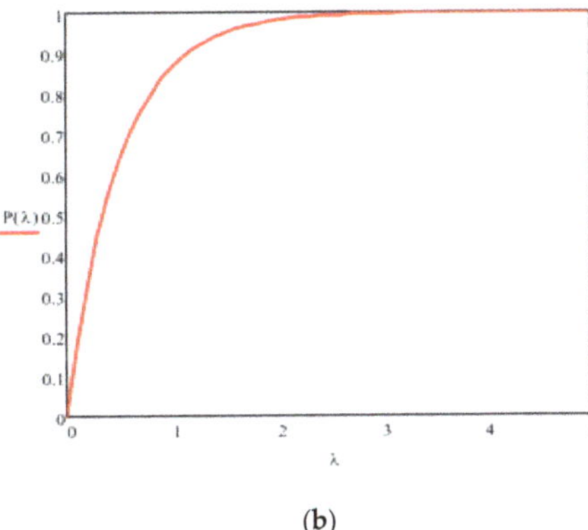

(b)

Figure 3. Dependence of (**a**) the average number of packets C that is successfully transmitted and (**b**) the probability of losses from collisions P from the intensity λ.

Thus, the average number of packets that are successfully transmitted during a time T, where the intensity λ can be determined as

$$C_{packets} = \lambda * P_{successful} = \lambda * \exp^{-2\lambda}, \tag{12}$$

where $C_{packets}$ refers to the average number of packets that are successfully transmitted. Figure 3 shows the effect of packets' arrival rate on the average number of successfully transmitted packets.

3.4. Calculation Gateway Capacity

Consider a network LoRa operating at a frequency of 868 MHz and a bandwidth of 125 kHz. The number of radio-frequency channels N_f is equal to 8. It is assumed that the nodes transmit a packet of 8 bytes of payload and a preamble with six symbols with a transmission rate of two packets per hour. The admissible probability of loss due to collisions is 2%. If two nodes or more transmit their packets simultaneously at the same SF spreading factor, a collision is likely to occur.

The total transmission time of one packet can be calculated by adding the total up-link and down-link times. The up-link time is the packet transmission time from the node to the gateway, while the down-link time is the transmission time of the confirmation packet from the gateway to the node. For a certain spreading factor, the total transmission time of a packet, T_{SF}, can be calculated as

$$T_{SF} = T_{SF-UL-pack} + T_{SF-DL-pack}, \tag{13}$$

where $T_{\text{SF-UL-pack}}$ is the total up-link time, and $T_{\text{SF-DL-pack}}$ is the total down-link time. The LoRa gateway's capacity can be defined as the total number of packets that the gateway serves per day. This may be represented by the throughput and can be calculated as

$$\text{Throughput} = N_f * \Sigma_{\text{SF}} P_{\text{SF}} * \frac{N_{\text{ENpack}} * 3600 * \lambda 2\%}{T_{\text{SF}}}, \tag{14}$$

where N_{ENpack} is the total number of packets transmitted by one end node per day, $\lambda 2\%$ is the rate of arrival of packets at a probability of packet loss P_{loss} of 2, N_f is the total number of radio channels deployed by the LoRa network, and P_{SF} is the of using certain SF. Based on the results of Figure 3, the value of $\lambda 2\%$ is equal to 0.01.

4. Problem Statement

More cluster members should be expected to result in more significant savings in end devices' network resources. However, the number of end devices in a cluster is limited by the CH throughput and traffic generated and their physical location relative to the CH. Due to the radio channel's peculiarities, the channel resource between the cluster member and the CH may be different for different members or CH and the characteristic of the expected channel quality for the cluster members.

We assume different scenarios and selection criteria choices, both cluster members and CH. For example, we can follow the maximum throughput, uniform distribution among the cluster members or the maximum cluster members, and consider or predict users' traffic intensity. To choose one or another scenario, we need to know the resulting solution's characteristics.

We will characterize the cluster by the throughput between the cluster members and CH and the achievable data transfer rate. We will analyze the throughput of cluster members with different laws of end device distribution. Different clustering methods can provide different solutions in terms of the distribution of end devices within the cluster.

5. System Description

Building IoT applications provide a high-quality environment due to the massive amount of data collected through many sensors. Such sensors installed in monitoring sites will collect and analyze information about the air level, soil and water pollution; noise level, the level of reservoirs and rivers. All the information received from these heterogeneous applications generates a single LoRa network to provide a high QoS for all traffic types within the LoRa network. The QoS is measured as the packet reception ratio (PRR) function and throughput of the LoRa network. Thus, we must characterize the throughput value made in terms of the QoS. We will consider the throughput as an objective metric among the network elements t_{ij}. In our study, we also considered the head node (HN) of the cluster has already been determined without considering the clustering methods. We assumed the HN communication zone is a disk with a radius R, centered at the CH location, as shown in Figure 4.

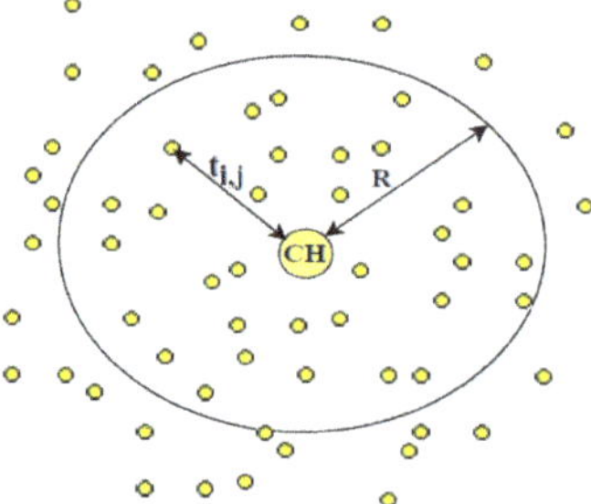

Figure 4. Considered cluster model.

We considered LoRa technology as a communication method among end nodes. We have to specify the nature of dependence t_{ij} on the network parameters. On the way of signal propagation to the receiver from the transmitter occurs environmental energy absorption, whereby the output signal at the receiver is significantly reduced, i.e., signal attenuation occurs. There are different attenuation models, taking into account various factors such as distance, carrier frequency, and obstacles in the path of signal propagation. One common attenuation model is described as

$$A(d) := 20 \log\left(\frac{\lambda}{4\pi d}\right)(dB).\tag{15}$$

The signal strength depends on the power of the transmitter. The majority of LoRa equipment has 25 mW transmitters. To describe the transmitter power, a relative value (power level) is often used, with the power of 1 mW, in decibels, defined as

$$P_{tx} := 10 \log 10\left(\frac{P}{10^{-3}}\right)(dBm),\tag{16}$$

where P is the transmitter power (W).

The signal power at the receiver input will be determined as follows:

$$P_{rx}(d) = P_{tx} - A(d),\tag{17}$$

where P_{rx} is the transmitter output signal power, and A(d) is the signal loss from the distance that can be calculated according to the formula (15). Using the data from Table 4 and Formula (17), we can plot a graph of the dependence of the data rate on the distance, as shown in Figure 5. The signal/noise + noise (SINR) and SNR values are the dependent data transmission rate on signal power.

Table 4. Dependence LoRa gateway's range on the receiver's sensitivity when using the ITU-RP.1238-5 attenuation model [28].

SF	6	7	8	9	10	11	12
RSSI (dBm)	−118	−123	−126	−129	−132	−134	−137
Bitrate (bit/s)	9375	5469	3125	1758	977	537	293
Distances (m)	270	333	410	506	623	716	824

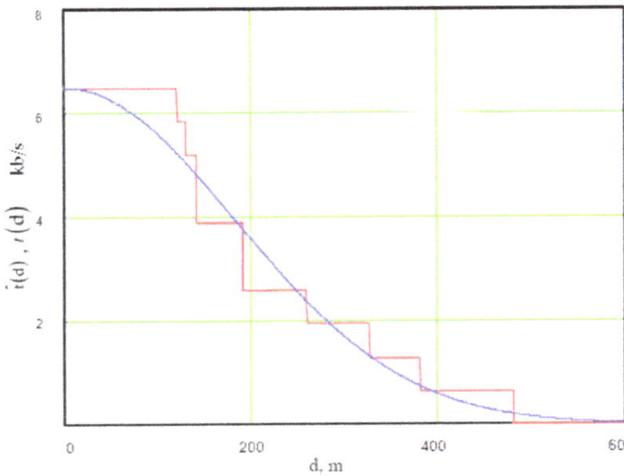

Figure 5. The dependence between the data transmission rate and distance.

Every parameter determined above impacts the channel throughput and can be chosen as a metric in a clustering task. In this work, we analyze throughput in a cluster.

In Figure 5, the throughput value depends on receiving and, in practice, has considerable dispersion.

In urban conditions for indoor application where there is a dense building, it is possible to apply the decay model ITU-RP.1238-5 for premises:

$$L(d) = 20\log(f) + N\log(d) + Lf(n) - 28, \tag{18}$$

where d is the distance in (m); n = 33—remote power loss factor; f is the center frequency of the signal (MHz); Lf (n) = 24 is the loss factor due to the passage of the signal through the obstacle (dB).

Considering the attenuation model, we depict the throughput dependency on the distance by the Gaussian function (Figure 5) [29].

$$\hat{t}(d) = \hat{t}(L(d)), \tag{19}$$

$$t(d) = \begin{cases} 0 & d < 0 \\ t_{max}\exp^{-\frac{d^2}{2c^2}} & 0 \le d \le R \\ 0 & d > R \end{cases}, \text{bit/s}, \tag{20}$$

where d is the distance (m); t_{max} is the maximum data transmission rate (bit/s); c is the curve half-width (m); R = arg{ $\hat{t}(d) = 0$ } (m) is the radius from CH to HN communication zone.

Since the throughput, according to the given model, depends on the distance, which to an arbitrarily chosen point should be considered as a random variable, then the throughput is a distribution function t can be calculated as

$$F(t) = \iint\limits_{D_t} f(x,y)dxdy, \tag{21}$$

where $f(x,y)$ refers to the users distribution function on a dick with radius R, D_t is the range of t values.

The probability density t is defined as

$$f(t) = \frac{dF(t)}{dt}. \tag{22}$$

The mathematical expected value t is

$$M(t) = \int_0^{t_0} t \cdot f(t)dt. \tag{23}$$

In our clustering model, we consider different types of user distributions, as discussed in Sections 5.1 and 5.2.

5.1. Uniform Distribution

We will presume that end nodes' uniform distribution is specified in the service area with a circle S and radius R (S = πR2). Thus, end nodes are distributed over the interval $0 \le r \le R$. The radius of the circle R can be defined as a solution to the following equation:

$$R = \arg\{ \hat{t}(d) = 0 \} (M). \tag{24}$$

The probability density function f(r), in this case, is constant and will be close to the uniform distribution for the distribution of end nodes inside the circle S:

$$f(r) = \frac{1}{S} - \frac{1}{\pi R^2}. \tag{25}$$

For the functional dependence of the throughput t on the distance between gateways has the form (20), that is

$$t(d) = t_{max} \exp^{-\frac{d^2}{2c^2}}. \tag{26}$$

We can express from (20) $d = c\sqrt{-2\ln\left(\frac{t}{t_{max}}\right)}$ (M).

According to (21), throughput distribution function F(t) on a circle with radius R will be determined as

$$F(t) = \int\limits_{0}^{2\pi} \int\limits_{c\sqrt{-2\ln\left(\frac{t}{t_{max}}\right)}}^{R} \tfrac{1}{S} r dr d\theta = \left. \tfrac{1}{2\pi R^2} r^2 \right|_{c\sqrt{-2\ln\left(\frac{t}{t_{max}}\right)}}^{R} \cdot 2\pi = \tag{27}$$

$$= \tfrac{1}{R^2}\left(R^2 + 2c^2\ln\left(\tfrac{t}{t_{max}}\right)\right) = 1 + \tfrac{2c^2}{R^2}\ln\left(\tfrac{t}{t_{max}}\right)$$

Based on (22), the probability density function of the throughput can be shown as

$$f(t) = \frac{dF_t(r)}{dr} = \frac{d}{dr}\left(1 + \frac{2c^2}{R^2}\ln\left(\frac{t}{t_{max}}\right)\right) = \frac{2c^2}{R^2 t}. \tag{28}$$

Probability density functions and throughput distribution are shown in Figure 6.

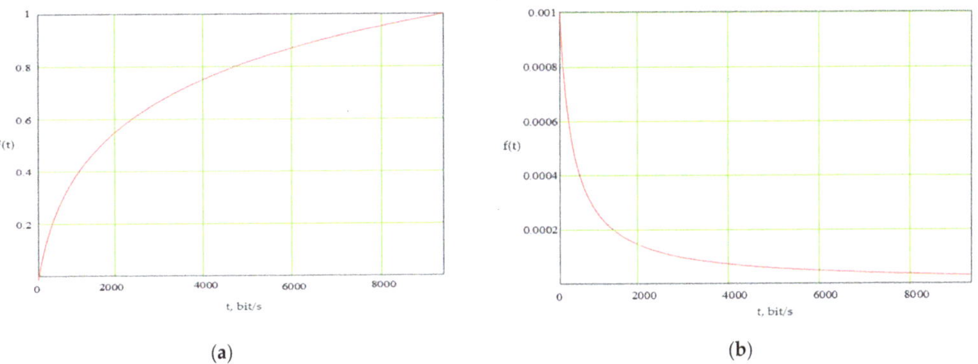

Figure 6. Probability of throughput distribution function F(t) (**a**) and throughput probability density function f(t) (**b**) for uniform distribution of nodes.

For the LoRa standard, the mathematical expectation of the throughput M (t) in the communication zone between end nodes according to (23) will be calculated as

$$M(t) = \int\limits_{t_{min}}^{t_{max}} t \cdot \frac{2c^2}{R^2 t} dt = 2 \cdot \frac{c^2}{R^2}(t_{max} - t_{min}) \text{ bit/s}. \tag{29}$$

When approximating the throughput function on the distance by using a uniform distribution of end nodes in the communication zone, the mathematical expectation of the throughput for the LoRa standard is 2654 bit/s.

5.2. Normal Distribution

Assuming that end nodes are randomly distributed in the communication area, then the distribution of end nodes over the communicated area is random, and a random value can describe their coordinates in each point on the surface, i.e., such a distribution can be given by a pair of random, independent coordinates x and y. Then the F(t) of the end nodes distribution can be defined as the joint distribution function of the random values x and y.

In this case, we will consider the normal distribution of end nodes with a scattering center at the center of the circle representing the communication area. Equation (30) defines the density distribution in both coordinates (x and y):

$$f(x,y) = \frac{1}{2\pi\sigma^2}\exp^{-\frac{x^2+y^2}{2\sigma^2}}, \text{where}\sqrt{x^2+y^2} = r, \tag{30}$$

where σ is the standard deviation (RMS).

For throughput analysis, we will assume that the probability of a point falling inside the circle of radius R is equal to 1, i.e., we will only consider users within the circle. The law of normal distribution is unlimited in both coordinates x and y. This assumption introduces a certain error; this error, a truncated normal distribution was used, which is bounded in both coordinates by the radius R. This distribution cannot be normal since it is bounded. The distribution of users within the communication area described by a circle S can be described by the "truncated normal" distribution [29].

$$f(x,y) = K(\sigma, R) \cdot \frac{1}{2\pi\sigma^2}\exp^{-\frac{x^2+y^2}{2\sigma^2}}, \tag{31}$$

where

$$K(\sigma,R) = \frac{1}{\iint\limits_{S_R} \frac{1}{2\pi\sigma^2}\cdot\exp^{-\frac{x^2+y^2}{2\sigma^2}}\,dxdy}. \tag{32}$$

Here S_R denotes the service area bounded by the circle with radius R.

Based on the obtained expressions, we find the F(t) inside the circle with radius R. For this, from (19) and (20), we obtain $d = c\sqrt{-2\ln\left(\frac{t}{t_{max}}\right)}$.

Then from (20), according to (31) and (32), the throughput distribution function can be determined as

$$F(t) = K(\sigma,R)\int\limits_0^{2\pi}\int\limits_{c\sqrt{-2\ln\left(\frac{t}{t_{max}}\right)}}^{R}\frac{1}{2\pi\sigma^2}\cdot\exp^{-\frac{r^2}{2\sigma^2}}\cdot r\,drd\theta =$$

$$= K(\sigma,R)\frac{1}{2\pi}\cdot\exp^{-\frac{r^2}{2\sigma^2}}2\pi\left|\begin{array}{c}R\\[2pt]c\sqrt{-2\ln\left(\frac{t}{t_{max}}\right)}\end{array}\right. = \tag{33}$$

$$= K(\sigma,R)\left(\left(\frac{t}{t_{max}}\right)^{\frac{c^2}{\sigma^2}} - \exp^{-\frac{R^2}{2\sigma^2}}\right).$$

Based on the obtained expression (33), the throughput probability density, according to (34), will be determined as

$$f(t) = K(\sigma,R)\frac{c^2}{\sigma^2 t_{max}}\left(\frac{t}{t_{max}}\right)^{\frac{c^2}{\sigma^2}-1}. \tag{34}$$

Figure 7 shows the probability density and throughput distribution function results.

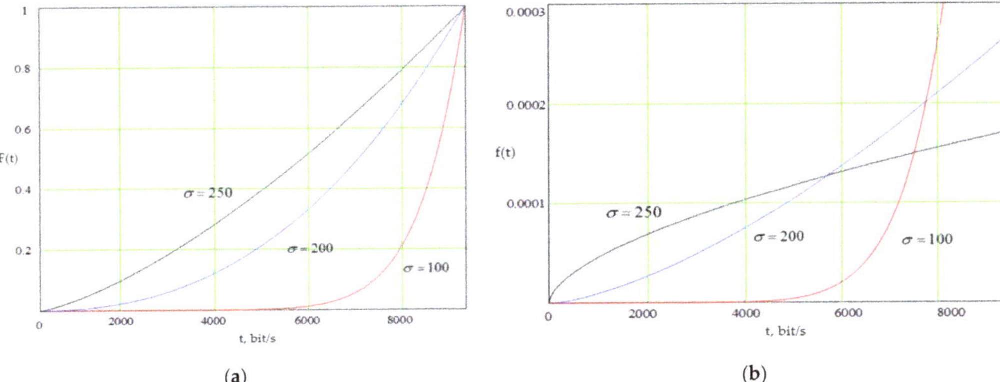

Figure 7. Probability distribution function F(t) (**a**) and probability density function f(t) (**b**) of the throughput (normal distribution of users).

From the expression for the throughput probability density in (34), an expression for the mathematical expectation of the throughput value can be defined as

$$M(t) = \int_0^{t_0} t \cdot f(t)dt = K(\sigma, R)\frac{c^2}{\sigma^2 + c^2}\left(t_{max} - t_{min}\left(\frac{t_{min}}{t_{max}}\right)^{\frac{c^2}{\sigma^2}}\right). \tag{35}$$

For the LoRa standard, the value of M (t) at σ of RMS 100, 200, 250 m takes the values of 8520, 6680, and 5750 bps, respectively.

6. Clustering Method Selection

We will consider a network of 20 thousand nodes in a field with 10 km sides. The coordinates of the nodes are distributed randomly according to a uniform distribution law. The maximum range of the lock is 824 m (from Table 4). We use two clustering methods [30] to the specifics of the projected communication network described by radio signal attenuation models and the distribution of subscriber traffic on the served area the previous section. The formation of a cluster consists of choosing a group of end devices and distributing their functionality within the cluster. The solution of the clustering problem is similar to the solution of the optimization problem in which some metric $d(m, p_m)$ is minimized (maximized), which characterizes the "distance" between a cluster member and the cluster center $p_m = \frac{1}{|c|}\sum_{m\in c} m$. Throughput, distance, and energy efficiency can act as such a metric. A dynamic programming method is used for the solution, which minimizes $d^2(m, p_m)$ across all clusters.

$$C = \min\sum_{c\in C}\sum_{m\in c} d^2(m, p_m). \tag{36}$$

For the formation, it is necessary to determine the method for finding an optimal solution. The QoS for traffic within the cluster between the CH and the network depends on the throughput of the channels between the cluster members and between the CH and LoRa Gateway and the traffic intensity generated by end devices.

Clustering algorithms K-means Algorithm 1 and FOREL Algorithm 2 are used to optimize communication network capacity.

$$\{x_i, y_i\} = \underset{x_i, y_i}{\operatorname{argmin}}\sum_{j=1}^{k}\sum_{r=1}^{n_j} d\left(C_j, e_{j,r}\right), \ i = 1\ldots k, \tag{37}$$

where $d\left(C_j, e_{j,r}\right)$ is the distance between the center of mass of the j-th cluster C_j and the r-th element of the j-th cluster $e_{j,r}$;

When interpreting K-means algorithm as an optimization problem, its objective function can be expressed as follows:

$$M = \sum_{i=1}^{k} \sum_{x_j \in S_i} \left(x_j - \mu_i\right)^2, \tag{38}$$

where k is the number of clusters;

S_i—defines a set of objects (elements) of the *i*-th cluster;

μ_i—point of the center of mass of the *i*-th cluster (coordinates of the point of the center of mass);

x_j—object of the *j*-th cluster (object coordinates).

Algorithm 1. K-means

Require: The k is a number of clusters, $C_1, C_2, \ldots .C_K$ points that corresponds to the devices, C_{MJ} j = 1,...,k—centers of clusters (mass centers).
Input: Set K random points $I = \{I_1, I_2, \ldots, I_n\}$ **Output:** Centers $(C_1, \ldots .C_K)$ C_{list} List of Clusters.
Procedure: Mode selection and K-Means clustering Algorithm.
Choose K initial centers $C_{MJ} = m_1 m_2, \ldots, m_k$.
For: $C_J <= C_{MJ}$ **do**
Set new centers of mass $\hat{m}_1$, $\hat{m}_2$, $\ldots$, $\hat{m}_k$ /*using Equations (39) or (40) */
If $\hat{m}_1$, $\hat{m}_2$, $\ldots$, $\hat{m}_k = m_1 m_2, \ldots, m_k$
Then
Set m_1 is new centers of mass /*using Equations (39) or (40) */
Each object X_i is assigned to the nearest C_i; for the resulting groups, the centers of mass are calculated.
Transition C_M $(C_I = C_M)$.
End for
fix C_j as the centers of the masses of the clusters, and X_i as the elements of the J cluster
End procedure.

Algorithm 2. FOREL

Require: The R is a communication rage (radius of the service area), the cluster number *i* = 1. C_1, $C_2, \ldots .C_K$ points that corresponds to the devices, C_{MJ} j = 1, ..., k—centers of clusters (mass centers).
Input: Set K random points $I = \{I_1, I_2, \ldots, I_n\}$.
Output: Centers $(C_1, \ldots C_K)$ C_{list} List of Clusters.
Procedure: Mode selection and FOREL clustering Algorithm.
Choose K initial centers $C_{MJ} = m_1 m_2, \ldots, m_k$.
For: True **do**
for all X_i points at a distance of $C_I <= R$ calculate the center of mass (C_M)/*using equations (39) or (40) */
while: $C_i = C_M$
Transition C_M $(C_I = C_M)$.
End while
fix C_j as the centers of the masses of the clusters, and X_i as the elements of the J cluster
End for
End procedure

The difference $\left(x_j - \mu_i\right)$ is the Euclidean distance between the cluster object and the center of the given cluster's mass.

When using K-means algorithm for two-dimensional space, i.e., when each of the objects has two characteristics (x and y) coordinates, each object can be considered a point on the flatness, characterized by its two coordinates (x_j, y_j).

The coordinates of the center mass of the j-th cluster is defined as the average value for each of the coordinates:

$$C_j = \left\{ x_j, y_j \right\}, \ x_j = \frac{1}{n_j} \sum_{r=1}^{n_j} x_{j,r}, \ y_j = \frac{1}{n_j} \sum_{r=1}^{n_j} y_{j,r}, \tag{39}$$

where n_j is the number of elements in the j-th cluster $x_{j,r}, y_{j,r}$ are the coordinates of the r-th element of the j-th cluster.

Along with coordinates, each object can be characterized by a certain numerical parameter m_j. Taking into account the last coordinates of the center mass of the cluster will be defined as the coordinates of the center of mass of a flat figure as

$$x_i^{(\mu)} = \frac{1}{m_i^{(\Sigma)}} \sum_{j=1}^{n_i} m_j x_j, \ y_i^{(\mu)} = \frac{1}{m_i^{(\Sigma)}} \sum_{j=1}^{n_i} m_j y_j, \ m_i^{(\Sigma)} = \sum_{j \in S_i} m_j. \tag{40}$$

The fundamental difference between the FOREL method and the K-means method is that the FOREL algorithm does not specify the number of clusters but assumes specifying the cluster size R.

When using this algorithm, expression (37) is minimized. The coordinates of the centers of mass of clusters are determined, which can be taken as positions for placing gateways.

7. Evaluation Results

Using the FOREL and K-means clustering methods, a network consisting of 25 clusters was obtained as a result of modeling. The simulation results for both clustering methods are illustrated in Figure 8a FOREL and Figure 8b K-means.

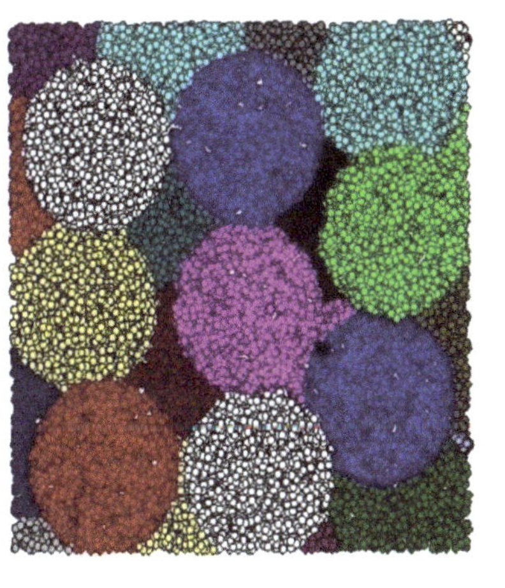

(a)   (b)

Figure 8. Simulation results for (a) FOREL and (b) K-means clustering methods.

To compare the two considered clustering methods for LoRa networks, we consider the distribution of cluster members relative to the cluster's center as the comparison metric. Distribution is obtained by simulation modeling. For center-of-mass, we used the following expression:

$$s = gm_j - Cm_i, i = 1 \ldots |C|, j = 1 \ldots n_i, m_j \in c_i, \tag{41}$$

where Cm_i, is the center-of-mass coordinate of cluster i; gm_j is the coordinate of m_j element of c_i cluster; c_i is the cluster i; gm_j is the element j of cluster i; n_i is the number of elements within the cluster i, and $|C|$ is the number of clusters.

It can be seen from the above results that the resulting clusters differ significantly in shape and size. This solution is quite obvious since the clustering problem with a high density of users was close to covering the service area with several circles (the minimum number of circles). Naturally, suppose the circles have the same radius. In that case, it is impossible to avoid their intersections, which ultimately leads to the formation of clusters of different shapes with different numbers of elements.

Figure 9 shows the relative distribution of nodes in the considered cluster for both considered clustering algorithms. The vertical axis in Figure 9 represents the number of nodes, while the horizontal axis indicates the relative distance between nodes and the cluster center. The negative value of the relative distance in Figure 9 shows that the nodes are located at the right side of the cluster center. As indicated in Figure 9, the main distribution of the nodes is located around the cluster center for both clustering methods.

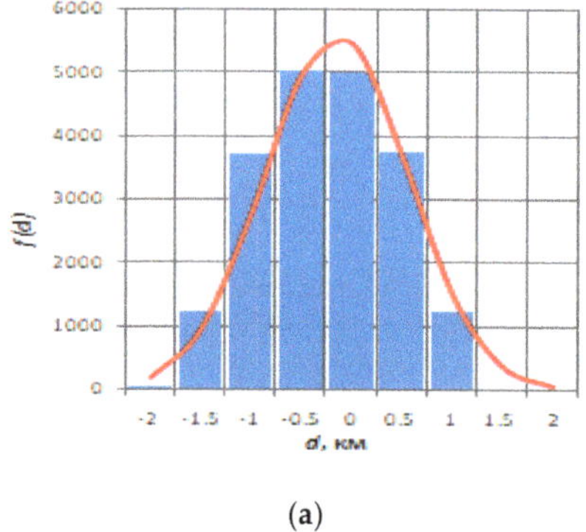

(a)

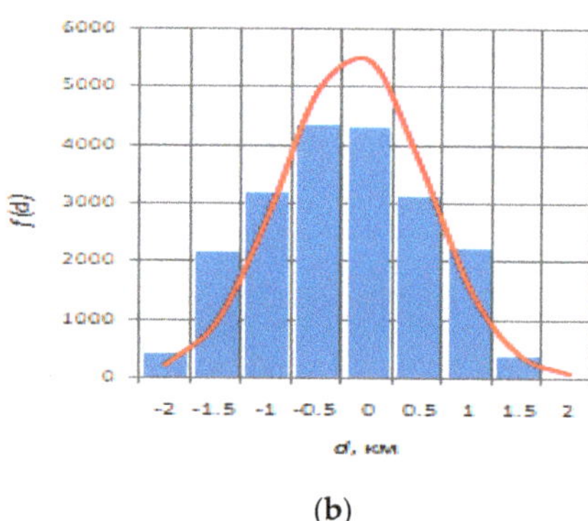

(b)

Figure 9. Relative distribution of nodes in the clusters for (**a**) FOREL clustering method and (**b**) K-mean clustering method.

Thus, it is possible to determine the average number of nodes in each SF zone; according to Table 4, Figure 10 indicates the average number of nodes in each SF zone for various clustering methods.

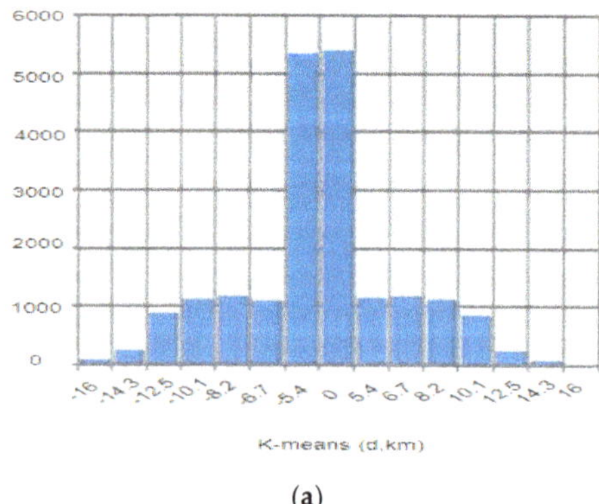

(a)

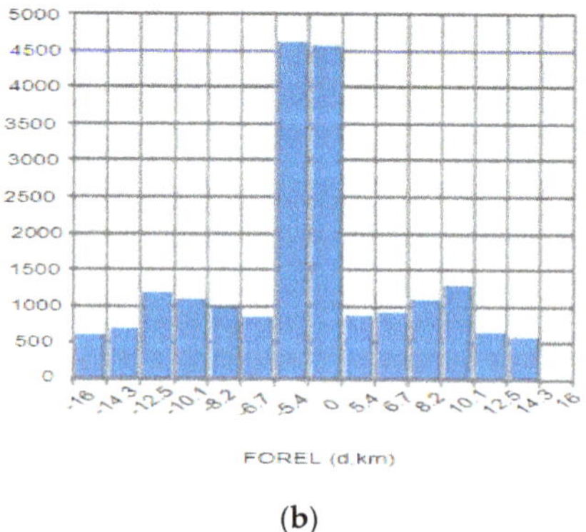

(b)

Figure 10. The average number of nodes in different SF zones (in percent, %) for (**a**) K-mean clustering method and (**b**) FOREL clustering method.

We calculate the capacity of the LoRa gateway for the two clustering methods discussed above, where the nodes are distributed differently by the area of the radio coverage zones. For FOREL = {45.96%; 8.54%; 9.47%; 11%; 12.34%; 6.7%; 5.99%} and for K-means {53.75%; 11.25%; 11.76%; 11.24%; 8.7%; 2.3%; 1.0%} as shown in Figure 11.

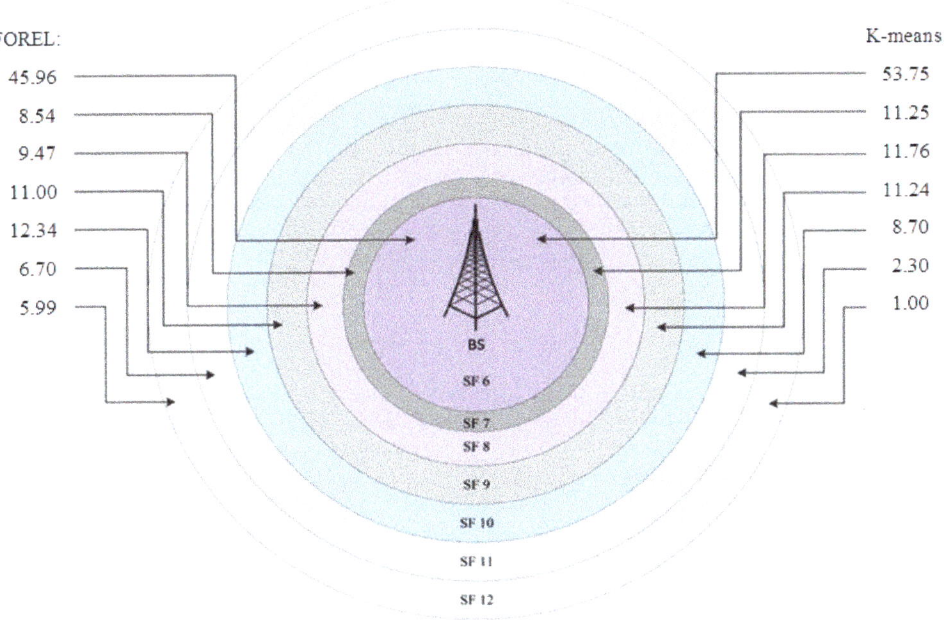

Figure 11. The average number of nodes in different SF zones for FOREL and K-means clustering methods (in percent, %).

The LoRa gateway's capacity can be calculated for FOREL and K-means clustering methods as shown in using Equation (14). Table 5 indicates this value and the total number of nodes.

Table 5. Calculation of the capacity of the LoRa gateway.

Clustering Method	NENpack (Per Day)	$\lambda 2\%$	Throughput	Number of Connected Devices to a Gateway
FOREL	24	0.01	79.223	3300
K-means	24	0.01	92.292	3845

Thus, to build a LoRa network that handles 20,000 end devices with the above-introduced characteristics, five to six gateways are needed to deploy the clustering algorithm, either FOREL or K-means.

The experiments showed that with a huge number of clusters, the K-means method allows obtaining results with a normal distribution of nodes across clusters and a higher throughput than the FOREL method. The network nodes are uniformly distributed in the service area.

By applying the above-proposed model, the average throughput value with a normal distribution is 8520 bit/s and with a uniform distribution is 2654 bit/s. Thus, to further assess the effectiveness of our solution, we compare our model with the literature. Table 6 describes a comparison of our proposed model with the literature for throughput values.

Table 6. Comparison of our proposed results with the literature for throughput values.

Existing Work	Throughput (t) bit/s Uniform Distribution	Throughput (t) bit/s Normal Distribution
[19]	2122	4522
[20]	2206	4653
[Proposed]	2654	8520

8. Discussion

Before obtaining the results achieved, clustering is to save network resources and improve the QoS. In both cases, the criterion for forming the cluster should take into account the available resources, transfer traffic from end nods, and QoS requirements.

The results presented in the previous section analysis of the channel capacity between the CM and the CH showed its dependence on the distribution of end nodes. In particular, the results showed that a larger average throughput value is achieved with a normal distribution than with a uniform distribution.

Cluster analysis algorithms can be applied to find partial solutions to selecting the coordinates of access points, considering the nature of the distribution of traffic sources over the served territory and CH nodes' choice.

Applying the algorithms of cluster analysis, necessary to the specifics of the projected communication network, which is described by models of radio signal attenuation and the model of distribution of subscriber traffic on the served area, thus the solutions, which are obtained in the previous section as a result of cluster analysis application, are the partial solutions of the optimization problem in communication network capacity.

Simulation of network clustering based on FOREL and K-means algorithm in the above section showed that these algorithms allow forming clusters from local groups of end nodes. The cluster size is given by the parameter R and the number of clusters selected, taking into account the availability and quality of communication.

To form clusters of LoRa end devices based on the bandwidth of the channels between the HN and CM can be used known methods of clustering objects. It is possible to use clustering methods based on finding the centroid, evaluating connectivity, density.

Our simulation results of two centroid clustering methods showed that the law of elements distribution in clusters depends on the number of clusters. With a relatively huge number of clusters, the service area's distribution elements are close to the normal law and relatively small to the uniform law.

The clustering method and parameters' choice is reflected in the distribution of cluster elements and the throughput between cluster elements and HN.

9. Conclusions and Future Work

This paper proposed a novel cluster throughput model to estimate the expected throughput value in clusters of end devices built using LoRa networks. In contrast to known models, it allows for the description of the throughput distribution function in a cluster of end devices.

We developed clustering algorithms using dense LoRa-based IoT networks that allow clustering of end devices according to the criterion of maximum served traffic. The algorithms are built based on two-common methods, K-means and Forel. In contrast to existing methods, the developed method provides the maximum value of served traffic in a cluster.

Results highlight our solution's effectiveness that our proposed model achieved a larger average throughput value with a normal distribution than with a uniform distribution.

Future work shall consider a novel clustering algorithm for achieving a higher level of flexibility so that the network will support the insertion and cutting of network devices and shall consider more QoS metrics. Furthermore, the LoRa network will have the flexibility level that enables the set-up of new applications.

Author Contributions: Conceptualization, M.S.A.M.; methodology, M.S.A.M., P.W. and M.W.; software, P.W., M.W. and A.R.; validation, M.S.A.M.; formal analysis, A.R.; investigation, M.S.A.M.; resources, P.W., M.W. and N.N.J.; data curation, M.S.A.M.; writing—original draft preparation, M.S.A.M.; writing—review and editing, A.R. and N.N.J.; visualization, M.S.A.M.; supervision, P.W. and M.W.; project administration, P.W. and M.W.; funding acquisition, P.W. and M.W. All authors have read and agreed to the published version of the manuscript.

Funding: This research was funded by The National Key Research and Development Program of China (2017YFE0123000).

Institutional Review Board Statement: Not applicable.

Informed Consent Statement: Not applicable.

Data Availability Statement: Data available on request due to restrictions–privacy. The data presented in this study are available on request from the corresponding author.

Acknowledgments: Many thanks are due to Ammar Muthanna and Min Wei for the assistance in the methodological work.

Conflicts of Interest: The authors declare no conflict of interest.

References

1. Rghioui, A.; Oumnad, A. Internet of things: Surveys for measuring human activities from everywhere. *Int. J. Electr. Comput. Eng.* **2017**, *7*, 2474–2482. [CrossRef]
2. Semtech, "Smart cities transformed using semtech's LoRa technology," July 2017, White Paper. Available online: https://info.semtech.com/smart_city_white_paper_download (accessed on 22 December 2020).
3. Khan, P.W.; Byun, Y.C.; Park, N. IoT-Blockchain Enabled Optimized Provenance System for Food Industry 4.0 Using Advanced Deep Learning. *Sensors* **2020**, *20*, 2990. [CrossRef] [PubMed]
4. Khan, P.W.; Byun, Y.A. Blockchain-Based Secure Image Encryption Scheme for the Industrial Internet of Things. *Entropy* **2020**, *22*, 175. [CrossRef] [PubMed]
5. Muthanna, M.S.A.; Lyachek, Y.T.; Musaeed, A.M.O.; Esmail, Y.A.H.; Adam, A.B.M. Smart system of a real-time pedestrian detection for smart city. In Proceedings of the IEEE Conference of Russian Young Researchers in Electrical and Electronic Engineering, Saint Petersburg and Moscow, Russia, 27–30 January 2020; pp. 45–50.
6. Semtech, LoRa Modulation Basics, AN1200.22 Revision 2. Semtech. 17 May 2015. Available online: https://www.semtech.com/uploads/documents/an1200.22.pdf (accessed on 22 December 2020).
7. Wydra, M.; Kubaczynski, P.; Mazur, K.; Ksiezopolski, B. Time-Aware Monitoring of Overhead Transmission Line Sag and Temperature with LoRa Communication. *Energies* **2019**, *12*, 505. [CrossRef]
8. Adil, N.; Khan, P.W.; Byun, Y.C. Performance Enhancement through Communication Offloading for Energy Efficiency on Mobile Cloud Computation. *Int. J. Sci. Technol.Res.* **2020**, *9*, 186–194.
9. Muthanna, M.S.A.; Wang, P.; Wei, M.; Abuarqoubet, A.; Alzu'bi, A.; Gull, H. Cognitive control models of multiple access IoT networks using LoRa technology. *Cognitive Syst. Res.* **2021**, *65*, 62–73. [CrossRef]
10. Adelantado, F.; Vilajosana, X.; Tuset-Peiro, P.; Martinez, B.; MeliaSegui, J.; Watteyne, T. Understanding the Limits of LoRaWAN. *IEEE Commun. Mag.* **2017**, *55*, 34–40. [CrossRef]
11. Bor, M.C.; Roedig, U.; Voigt, T.; Alonso, J.M. Do LoRa Low-Power Wide-Area Networks Scale? In Proceedings of the 19th ACM International Conference on Modeling, Analysis and Simulation of Wireless and Mobile Systems, New York, NY, USA, 13–17 November 2016; pp. 59–67.
12. Augustin, A.; Yi, J.; Clausen, T.; Townsley, W.M. A Study of LoRa: Long Range: Low Power Networks for the Internet of Things. *Sensors* **2016**, *16*, 1466. [CrossRef] [PubMed]
13. Liando, J.C.; Gamage, A.; Tengourtius, A.W.; Li, M. Known and unknown facts of lora: Experiences from a large-scale measurement study. *ACM Transactions Sens. Netw. (TOSN)* **2019**, *15*, 16. [CrossRef]
14. Muthanna, M.S.A.; Wang, P.; Wei, M.; Ateya, A.A.; Muthanna, A. *Toward an Ultra-Low Latency and Energy Efficient LoRaWAN*; Galinina, O., Andreev, S., Balandin, S., Koucheryavy, Y., Eds.; NEW2AN/ruSMART-2019. LNCS; Springer: Cham, Switzerland, 2019; Volume 11660, pp. 233–242. [CrossRef]
15. Luvisotto, M.; Tramarin, F.; Vangelista, L.; Vitturi, S. On the Use of LoRaWAN for Indoor Industrial IoT Applications. *Wirel. Commun. Mob. Comput.* **2018**, *2018*, 1–11. [CrossRef]
16. Bor, M.; Roedig, U. LoRa Transmission Parameter Selection. In Proceedings of the 2017 13th International Conference on Distributed Computing in Sensor Systems (DCOSS), Ottawa, ON, Canada, 5–7 June 2017; pp. 27–34.
17. Lavric, A. LoRa (Long-Range) High-Density Sensors for Internet of Things. *Sensors* **2019**. [CrossRef]
18. Sanchez-Iborra, R.; Liaño, I.G.; Simões, C.; Couñago, E.; Skarmeta, A. Tracking and Monitoring System Based on LoRa Technology for Lightweight Boats. *Electronics* **2018**, *8*, 15. [CrossRef]

19. Hauser, V.; Hégr, T. Proposal of Adaptive Data Rate Algorithm for LoRaWAN-Based Infrastructure. In Proceedings of the 2017 IEEE 5th International Conference on Future Internet of Things and Cloud (FiCloud), Prague, Czech Republic, 21–23 August 2017; pp. 85–90. [CrossRef]
20. Martinez-Sandoval, R.; García-Sánchez, A.; García-Haro, J. Performance optimization of LoRa nodes for the future smart city/industry. *EURASIP J. Wirel. Commun. Netw.* **2019**, 1–13. Available online: https://doi.org/10.1186/s13638-019-1522-1 (accessed on 22 December 2020).
21. AN1200.22 LoRaTM Modulation Basics, Revision 2, Semtech Corporation. May 2015. Available online: http://www.semtech.com/images/datasheet/an1200.22.pdf (accessed on 15 August 2020).
22. LoRa Alliance. *LoRaWAN Specification, V. 1.1*; LoRa Alliance: Fremont, CA, USA, 2017. Available online: https://lora-alliance.org/resource-hub/lorawanr-specification-v11 (accessed on 15 September 2019).
23. Vangelista, L. Frequency Shift Chirp Modulation: The LoRa Modulation. *IEEE Signal Process. Let.* **2017**, *24*, 1818–1821. [CrossRef]
24. Semtech, SX1272/73 Low Power Long Range Transceive. Available online: https://www.semtech.com/uploads/documents/sx1272.pdf (accessed on 22 December 2020).
25. Baccelli, F.; Blaszczyszyn, B.; Muhlethaler, P. An Aloha protocol for multihop mobile wireless networks. *IEEE Trans. Inf. Theory* **2006**, *52*, 421–436. [CrossRef]
26. Yasmin, R.; Petäjäjärvi, J.; Mikhaylov, K.; Pouttu, A. On the integration of LoRaWAN with the 5G test network. In Proceedings of the 2017 IEEE 28th Annual International Symposium on Personal, Indoor, and Mobile Radio Communications (PIMRC), Montreal, QC, Canada, 8–13 October 2017; pp. 1–6.
27. Blaszczyszyn, B.; Muhlethaler, P. Stochastic analysis of nonslotted Aloha in wireless ad-hoc networks. In Proceedings of the IEEE INFOCOM, San Diego, CA, USA, 14–19 March 2010; pp. 1–9.
28. ITU-R Recommendations. *Propagation Data and Prediction Methods for the Planning of Indoor Radio Communication Systems and Radio Local Area Networks in the Frequency Range 900MHz to 100GHz, ITU-R P.1238-5*; International Telecommunication Union: Geneva, Switzerland, 2007.
29. Weisstein, E.W. Gaussian Function. Available online: https://mathworld.wolfram.com/ (accessed on 27 November 2002).
30. Boyinbode, O.; Le, H.; Mbogho, A.; Takizawa, M.; Poliah, R. A Survey on Clustering Algorithms for Wireless Sensor Networks. In Proceedings of the 2010 13th International Conference on Network-Based Information Systems, Takayama, Japan, 14–16 September 2010; pp. 358–364. [CrossRef]

 *information*

Article

Time-Optimal Gathering under Limited Visibility with One-Axis Agreement [†]

Pavan Poudel and Gokarna Sharma *

Department of Computer Science, Kent State University, Kent, OH 44240, USA; ppoudel@cs.kent.edu
* Correspondence: sharma@cs.kent.edu
† A preliminary version of this article has been published in SSS'17 conference.

Abstract: We consider the distributed setting of N autonomous mobile robots that operate in *Look-Compute-Move* (LCM) cycles following the well-celebrated classic oblivious robots model. We study the fundamental problem of gathering N autonomous robots on a plane, which requires all robots to meet at a single point (or to position within a small area) that is not known beforehand. We consider limited visibility under which robots are only able to see other robots up to a constant Euclidean distance and focus on the time complexity of gathering by robots under limited visibility. There exists an $\mathcal{O}(D_G)$ time algorithm for this problem in the fully synchronous setting, assuming that the robots agree on one coordinate axis (say north), where D_G is the diameter of the visibility graph of the initial configuration. In this article, we provide the first $\mathcal{O}(D_E)$ time algorithm for this problem in the asynchronous setting under the same assumption of robots' agreement with one coordinate axis, where D_E is the Euclidean distance between farthest-pair of robots in the initial configuration. The runtime of our algorithm is a significant improvement since for any initial configuration of $N \geq 1$ robots, $D_E \leq D_G$, and there exist initial configurations for which D_G can be quadratic on D_E, i.e., $D_G = \Theta(D_E^2)$. Moreover, our algorithm is asymptotically time-optimal since the trivial time lower bound for this problem is $\Omega(D_E)$.

Keywords: distributed algorithms; mobile robots; classic oblivious robot model; gathering; time complexity; visibility; connectivity

Citation: Poudel, P.; Sharma, G. Time-Optimal Gathering under Limited Visibility with One-Axis Agreement. *Information* **2021**, *12*, 448. https://doi.org/10.3390/info12110448

Academic Editor: Giovanni Viglietta

Received: 30 August 2021
Accepted: 24 October 2021
Published: 27 October 2021

Publisher's Note: MDPI stays neutral with regard to jurisdictional claims in published maps and institutional affiliations.

1. Introduction

In the classic model of distributed computation by mobile robots, also known as the $OBLOT$ model, each robot is modeled as a point in the plane [1,2]. The robots are *autonomous* (no external control), *anonymous* (no unique identifiers), *indistinguishable* (no external identifiers), *disoriented* (no agreement on local coordinate systems and units of distance measures), *oblivious* (no memory of past computation), and *silent* (no direct communication and actions are coordinated via only vision and mobility). They execute the same algorithm. Each robot proceeds in *Look-Compute-Move* (LCM) cycles: when a robot becomes active, it first obtains a snapshot of its surroundings (*Look*), then computes a destination based on the snapshot (*Compute*), and then finally moves towards the destination (*Move*) [2].

We consider the *gathering* problem in the $OBLOT$ model, where starting from any arbitrary (yet connected) initial configuration, all robots are required to meet at a single point (or to position within a small area) that is not known beforehand. Relaxing the requirement to meet at a single point by positioning them within a small area is performed to circumvent the impossibility result of gathering to a point in the asynchronous setting, even for two robots [3]. In fact, the algorithm we designed in this article positions all robots either at a single point not known beforehand or within a unit line segment not known beforehand depending on different conditions. Gathering is one of the most fundamental tasks and a central benchmark problem in distributed mobile robotics [4]. Early studies

33

on gathering in the *OBLOT* model solved it under *unlimited visibility*, where each robot is assumed to observe (the locations of) all other robots [5], and all the robots are connected to each other. The *viewing range* defines the maximum possible distance up to which a robot can observe other robots, and the *connectivity range* defines the maximum possible distance between any two nodes to be connected (i.e., to have an edge between them). Flocchini et al. [6] provided the first algorithm for gathering in the *OBLOT* model under *limited visibility*, where each robot can observe (the locations of) other robots within a fixed unit distance (*viewing range*), and each robot is connected to all other robots within that fixed unit distance (*connectivity range*), i.e., the viewing and connectivity ranges are the same. Subsequently, several algorithms were studied for this problem under different constraints [2,7–10]. These studies proved the correctness of the algorithms but provided no runtime analysis (except a proof of finite time termination).

The runtime analysis for gathering has been studied relatively recently [11–15]. Degener et al. [11] provided the first algorithm for this problem with runtime $\mathcal{O}(N^2)$ in expectation in the fully synchronous setting, where N is the total number of robots. Degener et al. [12] provided an $\mathcal{O}(N^2)$-time algorithm for this problem in a fully synchronous setting. They also showed that, for some initial configurations, their algorithm is essentially tight by providing a matching lower bound of $\Omega(N^2)$. Kempkes et al. [13] provided an $\mathcal{O}(OPT \log OPT)$-time algorithm for this problem under a slightly different continuous time setting, where OPT is the runtime of an optimal algorithm. All the above algorithms assume that both the viewing and connectivity ranges are of (fixed) radius 1. Recently, Cord-Landwehr et al. [14] provided an $\mathcal{O}(N)$-time algorithm for this problem for robots positioned on a grid in a fully synchronous setting. In this algorithm, it is assumed that robots have the viewing range of (distance) 20, i.e., each robot can observe other robots within a fixed distance of 20, but the connectivity range is one, i.e., two robots are connected if and only if they are vertical or horizontal neighbors on the grid. Moreover, each robot is assumed to possess memory for remembering a constant number of previous cycles. Recently, Fischer et al. [15] provided an $\mathcal{O}(N^2)$-time algorithm for gathering on a grid in the fully synchronous setting, if the memory is not available, by using the improved viewing range of 7.

The intriguing open question is whether a time-optimal algorithm can be designed for gathering under limited visibility and if possible, under what conditions. We define time optimality as follows: Let G be the visibility graph of an arbitrary initial configuration I of $N \geq 1$ robots in a plane. The robots in the system act as nodes of G. There is an edge between any two nodes in G if the distance between these two nodes is at most the connectivity range. Note that, according to the definitions above, the viewing and connectivity ranges may or may not be the same. If each robot is connected to all robots within its viewing range, then the viewing range also serves as the connectivity range; otherwise, the connectivity range is different than the viewing range. In order to solve the gathering problem, G must be connected [2]; G is connected if the robots (or nodes of G) cannot be separated into two subsets such that no robot of the one subset is connected to any robot of the other subset (and vice versa). For example, the authors in [14] used the viewing range 20, but the robots in horizontal or vertical distance of one are connected. Let D_G be the diameter of G, which is the greatest distance between any pair of nodes in G following the edges of G. Let D_E be the diameter of the initial configuration I, which is the greatest Euclidean distance between any pair of robots in I. Notice that for any I, $D_E \leq D_G$, and for some configurations the gap between D_G and D_E can be as much as quadratic on D_E, i.e., $D_G = \Theta(D_E^2)$. Figure 1 illustrates these ideas. Therefore, an $\mathcal{O}(D_E)$-time algorithm would be time-optimal for gathering starting from any initial connected configuration, since $\Omega(D_E)$ is the trivial time lower bound for robots to meet at a single point (or to position within a small area) starting from any arbitrary initial configuration. Hence, the open question specifically is whether an $\mathcal{O}(D_E)$-time algorithm can be designed for gathering for classic oblivious robots under limited visibility.

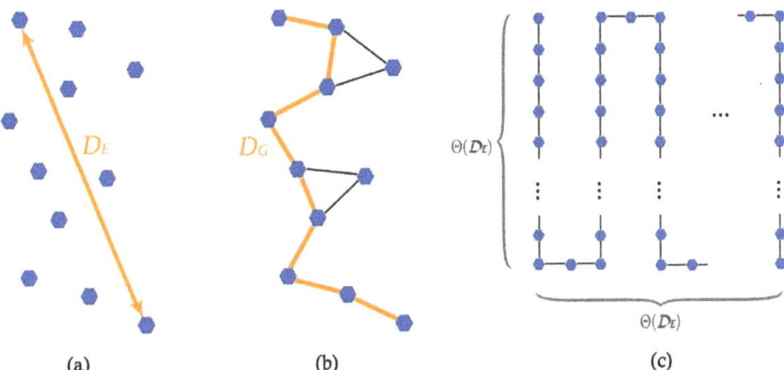

Figure 1. An illustration of two initial configuration dependent parameters, D_E (the Euclidean diameter) and D_G (the visibility graph diameter), and the relation between them: (**a**) the diameter D_E for an initial arbitrary configuration, (**b**) the visibility graph G with diameter D_G for the configuration of the left, and (**c**) an initial configuration showing the quadratic difference between D_E and D_G with $D_G = \Theta(D_E^2)$.

Recently, Izumi et al. [4] made progress towards addressing this open question. Specifically, they presented an $\mathcal{O}(D_G)$-time algorithm for gathering on the plane in a fully synchronous setting under limited visibility with the condition that robots agree on one coordinate axis. They used viewing range of one with an assumption that the visibility graph G remains connected even if the edges with the corresponding distance of greater than $1 - \frac{1}{\sqrt{2}}$ are removed from it. The assumption on the visibility graph G in Izumi et al. [4] essentially means that the connectivity range is of radius $1 - \frac{1}{\sqrt{2}}$ (different and in fact smaller than the viewing range of one).

There is still a large gap between the $\mathcal{O}(D_G)$ time bound of Izumi et al. [4] and the asymptotically optimal $\mathcal{O}(D_E)$ time bound, since D_G can be quadratic on D_E (Figure 1). This work closes this gap under the same one axis agreement with a slightly modified viewing range of $\sqrt{10}$ and the square connectivity range (if we do not explicitly write "square", then the viewing and connectivity ranges are circular) of $\sqrt{2}$ compared to the viewing range of one and the (circular) connectivity range of $1 - \frac{1}{\sqrt{2}}$ in [4] (if we consider the viewing range of one similar to [4], we need the square connectivity range of $1 - \frac{\sqrt{2}}{\sqrt{10}}$, and our algorithm again achieves $\mathcal{O}(D_E)$ runtime). Notice that the *square connectivity range* of distance c means that a robot is connected to all other robots inside or on the boundary of the (axis-aligned) square area with the (diagonal) distance from the robot to each corner of the square c. Notice also that the square connectivity range of $\sqrt{2}$ for a robot is equivalent to the L_∞-distance of 1 around the robot. Therefore, if we have both the viewing and connectivity ranges of c, then the area they enclose differs if the connectivity range is "square"; otherwise, they enclose the same area. Moreover, in contrast to Izumi et al. [4], which works in the fully synchronous setting, our algorithm works in the asynchronous setting. The algorithm presented by Izumi et al. [4] follows the movements of robots in one direction (either north or east) along D_G in such a way that in each round, starting from the southmost and westmost robots, each robot moves towards the farthest neighbor within its connectivity range. In our algorithm, robots do not follow D_G; instead, they gather at a point (or within a small area) that is not known beforehand in $O(D_E)$ rounds. Particularly, in our algorithm, all the robots gather at a single point not known beforehand under both axis agreements and inside a horizontal line segment of length one that is not known beforehand under one axis agreement. A preliminary version of this article has been published in SSS'17 conference [16], and this article extends that version by including many details and proofs that were missing in that version.

Contributions. We focus, in this article, on optimizing runtime for gathering under limited visibility. We consider autonomous, anonymous, indistinguishable, oblivious, and silent point robots (also called *swarms*) as in the classic $OBLOT$ model [2]. Robots agree on the unit of distance measure. The viewing range is $\sqrt{10}$—a robot can see all other robots within the fixed radius of at most distance $\sqrt{10}$. The square connectivity range is $\sqrt{2}$—a robot is connected to all other robots inside or on the boundary of the (axis-aligned) 2×2-sized square area for which its center is the position of the robot. In a LCM cycle, a robot can move to any position inside or on the square area, including its four corners. The challenge here is that robot movements must not harm swarm connectivity. As in Izumi et al. [4], we assume that robots agree on one coordinate axis (say north), but they may not agree on the other coordinate axis. Moreover, we assume that the robot setting is *asynchronous*—there is no notion of common time, and robots perform their LCM cycles arbitrarily. Furthermore, we assume that the robot moves are *rigid*—a robot in motion in each cycle cannot be stopped (by an adversary) before it reaches its destination at that cycle. Additionally, all previous algorithms assume that when two or more robots move to the same location, they are merged and act as a single robot. In this article, we do not have that assumption; in other words, even if robots are located at the same position and activated at different time, the gathering progress is achieved through the (individual) moves of those robots.

In this article, we prove the following result which, to our best knowledge, is the first algorithm for gathering that is asymptotically time-optimal for classic oblivious robots under limited visibility since the trivial time lower bound for gathering under limited visibility starting from any initial configuration of $N \geq 1$ robots is $\Omega(D_E)$.

Theorem 1. *For any initial connected configuration of $N \geq 1$ robots with the viewing range of $\sqrt{10}$ and the square connectivity range of $\sqrt{2}$ on a plane, gathering can be solved in $\mathcal{O}(D_E)$ time in the asynchronous setting, when robots agree on one coordinate axis.*

Notice that, the visibility graph G must be connected, since gathering may not be solvable under limited visibility if G is not connected [2,6]. Our selection of the viewing and (square) connectivity ranges and the assumption of one-axis agreement play an important role in proving Theorem 1. For both viewing and (circular or square) connectivity ranges of one, we conjecture that there is no $\mathcal{O}(D_E)$-time algorithm for gathering of classic oblivious robots in the asynchronous setting, even when robots agree on both coordinate axes. This is because a robot cannot move more than distance one in each LCM cycle to preserve connectivity, and only the end robots can move in each cycle. Therefore, if the robots are connected as shown in Figure 1, $O(D_G)$ time is required to gather them since only end robots can move and the rest cannot. For the viewing and (circular or square) connectivity ranges of constant > 1, we conjecture that there is no $\mathcal{O}(D_E)$-time algorithm for gathering of classic oblivious robots if the robots do not agree on any coordinate axis. This is because the robots' movements become arbitrary as there is no agreement on the coordinate axes. Thus, robots can only gather if they move following the diameter D_G, which only provides an $O(D_G)$-time algorithm.

Technique. Let L be the topmost horizontal line so that all the robots of any initial configuration I are either on the positions of line L or south from L. Let L' be the line parallel to L at distance one south of L. The main idea behind the algorithm is to make robots of I in the north of L' move to the positions of L' or south of L' in $\mathcal{O}(1)$ epochs, even under the asynchronous setting, where an epoch is the time interval for all N robots to execute their LCM cycle at least once (the formal definition of epoch is in Section 2). To accomplish this, we classify the moves of robots into three categories: *diagonal hops*, *horizontal hops*, and *vertical hops*. We will show that if all the robots in the north of L' make diagonal or vertical hops, they reach L' or south of L' in one epoch. However, if some of those robots make a horizontal hop, then in two epochs, they reach positions of L' or south of L' through the subsequent vertical or diagonal hop. We also show that, if some robots in the north of L' do not move in the first epoch, then they reach positions of L' or south of L' through the vertical or diagonal hop by the next two epochs.

Similarly, let L_b be the bottommost horizontal line (parallel to L) so that the robots of I are either on L_b or north of L_b. The main idea is to show that the robots on L_b do not move south of L_b forever. Specifically, we show that robots on L_b wait for all the robots in the north of L_b so that they meet at distance (at most) D south of L_b where D is proportional to the horizontal diameter of the initial configuration I. This is achieved by asking robots not to make any diagonal, horizontal, or vertical hops if they see at least a robot in the north at vertical distance 1 (or more) from their positions (i.e., on or outside the connectivity range of the corresponding robot).

Other Related Work. The classic oblivious robots model or the *OBLOT* model has been considered heavily in order to solve a diverse set of problems, such as scattering, gathering, convergence, circle formation, flocking, etc., in distributed mobile robotics. A comprehensive description of the state-of-the-art research on distributed computing by mobile robots can be found in these excellent books [2,17]. Much work on the classic model on these problems does not provide runtime analysis, for example, see papers on gathering in a non-predefined point [5–7,18,19]. Pagli et al. [20] considered gathering classic robots to a small area by avoiding collisions between robots. However, they also do not provide runtime analysis. Kirkpatrick et al. [19] studied gathering as a point convergence problem where starting from an arbitrary initial configuration, robots move in such a manner that they reach inside a circle of radius that is at most ϵ, $\epsilon > 0$. They proposed an algorithm to solve the problem in the k-asynchronous model (i.e., the degree of asynchrony is bounded to k); however, they showed that point convergence is unsolvable in the fully asynchronous model. In this article, we present an algorithm with runtime analysis to solve the gathering problem in the fully asynchronous model by assuming robots agree on one coordinate axis. Bhagat et al. [21] studied the limited visibility model for robots and presented different geometric pattern formation problems under limited visibility.

Gathering on a predefined point has been studied in several papers [22–24]. These papers studied gathering in the context of robots with an extent (i.e., *fat robots*). Applying these algorithms to the classic model solves gathering in $\mathcal{O}(D)$ time (provided that gathering point is known to robots), where D is the largest distance from any robot to the predefined gathering point. However, the runtime bound is provided only for the grid, and the gathering point is known to robots a priori. Recently, Braun et al. [25] studied the gathering problem in a three-dimensional Euclidean space under limited visibility and presented $O(n^2)$-time and $O(D_E \cdot n^{\frac{3}{2}})$-time algorithms in the fully synchronous and continuous time models, respectively.

The question of gathering on graphs instead of gathering in the plane was studied in [26–28]. Di Stefano and Navarra [27] assumed unlimited visibility and an asynchronous setting and proved the optimal bounds on the number of robot movements for special graph topologies such as trees and rings. D'Angelo et al. [28] showed that gathering can be solved in grids without multiplicity detection. Di Stefano and Navarra [26] extended the results of [28] to infinite grids and bounded the total number of robot movements.

Gathering is solved by circumventing the impossibility of gathering at a single point in some recent papers. The relaxation is on the gathering requirement: Gathering occurs within a small area instead of at a point. A prominent paper that solves gathering in a small area is written by Cord-Landwehr et al. [14] in which, starting from any arbitrary configuration on a grid, robots gathered within a 2×2-sized grid area. Cord-Landwehr et al. [29] provided an $\mathcal{O}(N)$-time algorithm for the robot convergence problem (converging toward a single not predefined point) [30].

Izumi et al. [31] considered the robot scattering problem (opposite of gathering) in the semi-synchronous setting and provided an expected $\mathcal{O}(\min\{N, D^2 + \log N\})$-time algorithm; here, D is the diameter of the initial configuration.

All the previous algorithms, including Izumi et al. [4], work in the fully synchronous setting, except for [11] which works in the one-by-one activation setting (also known as sequential activation). Our algorithm works in the asynchronous setting. Furthermore, all previous algorithms assume that when two or more robots move to the same location, they are merged as only one robot. Our algorithm does not merge robots; in other words, even if robots located at the same position are activated at different times, the gathering progress is achieved through the (individual) moves of those robots.

Roadmap. In Section 2, we detail the model and touch on some preliminaries. For the sake of simplicity in discussion, we first provide an $\mathcal{O}(D_E)$-time algorithm for robots on a grid agreeing on both the coordinate axes in Section 3. We then provide an $\mathcal{O}(D_E)$-algorithm for robots on a plane agreeing on both the coordinate axes in Section 4. In Section 5, we discuss how the algorithms of Sections 3 and 4 can be modified to solve gathering when robots agree on only one axis. Finally, we provide concluding remarks in Section 6 with a brief discussion.

2. Model and Preliminaries

Robots. We consider a distributed system of N robots (agents) from a set $\mathcal{Q} = \{r_0, r_1, \cdots, r_{N-1}\}$. Each robot is a (dimensionless) point that can move in an infinite two-dimensional real space $\mathbb{R}^2$. Throughout this article, we will use a point to refer to a robot as well as its position. We denote by $\text{dist}(r_i, r_j)$ the distance between two robots $r_i, r_j \in \mathcal{Q}$. Each robot r_i works under limited visibility and viewing range of each robot is $\sqrt{10}$, i.e., a robot r_i can see and be visible to another robot r_j if and only if $\text{dist}(r_i, r_j) \leq \sqrt{10}$. For some cases, e.g., for grid, the viewing range smaller than $\sqrt{10}$ is sufficient. We describe what exactly is the viewing range when we describe algorithms in Sections 3 and 5. The connectivity range of each robot is $\sqrt{2}$ following square connectivity, i.e., two robots have an edge between them on G if one robot is inside the (axis-aligned) 2×2-sized square area formed by the other robot being at its center. The robots agree on the unit of distance measure, i.e., the viewing and connectivity ranges of $\sqrt{10}$ and $\sqrt{2}$ are the same for each robot $r_i \in \mathcal{Q}$. The robots also agree on one coordinate axis, north (the assumption of robots agreeing on east is analogous). For the sake of simplicity in discussion, the algorithms in Sections 3 and 4 assume that robots agree on both coordinate axes. The assumption on both axis agreement is lifted in Section 5.

Look-Compute-Move. Each robot r_i is either active or inactive. When a robot r_i becomes active, it performs the "Look-Compute-Move" cycle as follows:

- *Look:* For each robot r_j that is within the viewing range of r_i, r_i can observe the position of r_j on the plane. Robot r_i also knows its own position;
- *Compute:* In any cycle, robot r_i may perform an arbitrary computation using only the positions observed during the "look" portion of that cycle. This includes determination of a (possibly) new position for r_i for the start of next cycle;
- *Move:* At the end of the cycle, robot r_i moves to its new position.

Robot Activation. In the fully synchronous setting ($\mathcal{FSYNC}$), every robot is active in every LCM cycle. In the semi-synchronous setting ($\mathcal{SSYNC}$), at least one robot is active, and over an infinite number of LCM cycles, every robot is often infinitely active. In the asynchronous setting ($\mathcal{ASYNC}$), there is no common notion of time, and no assumption is made on the number and frequency of LCM cycles in which a robot can be active. The only guarantee is that every robot is active infinitely often. Complying with the $\mathcal{ASYNC}$ setting, we assume that a robot "wakes up" and performs its *Look* phase at an instant of time. We also assume that during the *Move* phase, it moves in a straight line and stops only after reaching its destination point; in other words, the moves are rigid [2].

Runtime. For the $\mathcal{FSYNC}$ setting, time is measured in rounds. Since a robot in the $\mathcal{SSYNC}$ and $\mathcal{ASYNC}$ settings could stay inactive for an indeterminate interval of time, we bound a robot's inactivity and use the standard notion of epoch to measure runtime. An *epoch* is the smallest interval of time within which each robot is guaranteed to execute its LCM cycle at least once. Therefore, for the $\mathcal{FSYNC}$ setting, a round is an epoch. We will use the term "time" generally to mean rounds for the $\mathcal{FSYNC}$ setting and epochs for the $\mathcal{SSYNC}$ and $\mathcal{ASYNC}$ settings.

Square Area. Let $r_i \in \mathcal{Q}$ be a robot positioned at coordinate (x_i, y_i). Let L_i, L_i', respectively, be the horizontal and vertical lines passing through r_i. Since r_i knows north, r_i can easily compute L_i, L_i'. The *square area* for r_i, denoted as $SQ(r_i)$, is an area of the plane enclosed by four lines $L_{i,t}, L_{i,b}, L_{i,l}, L_{i,r}$ with $L_{i,t}, L_{i,b}$ being parallel to L_i (perpendicular to L_i') and passes through coordinates $(x_i, y_i + 1)$ and $(x_i, y_i - 1)$, respectively, and $L_{i,l}, L_{i,r}$ is perpendicular to L_i (parallel to L_i') and passes through coordinates $(x_i - 1, y_i)$ and $(x_i + 1, y_i)$, respectively. Notice that $SQ(r_i)$ is axis-aligned, and both the height and width of it is two. We denote by $p_{tl}, p_{bl}, p_{br}, p_{tr}$ the intersection points of lines $L_{i,t}$ and $L_{i,l}$, $L_{i,b}$ and $L_{i,l}$, $L_{i,b}$ and $L_{i,r}$, and $L_{i,t}$ and $L_{i,r}$, respectively. We can divide $SQ(r_i)$ to four quadrant squares $SQ_1(r_i), SQ_2(r_i), SQ_3(r_i)$, and $SQ_4(r_i)$ with both heights and widths as one. Let $SQ_1(r_i)$ and $SQ_2(r_i)$ be at the north of L_i and $SQ_3(r_i)$ and $SQ_4(r_i)$ be at the south of L_i. Moreover, let $SQ_1(r_i)$ and $SQ_3(r_i)$ be at west of L_i' and $SQ_2(r_i)$ and $SQ_4(r_i)$ be at east of L_i'. We say that the positions of L_i in $SQ(r_i)$ belong to $SQ_3(r_i)$ and $SQ_4(r_i)$. Figure 2a illustrates these ideas.

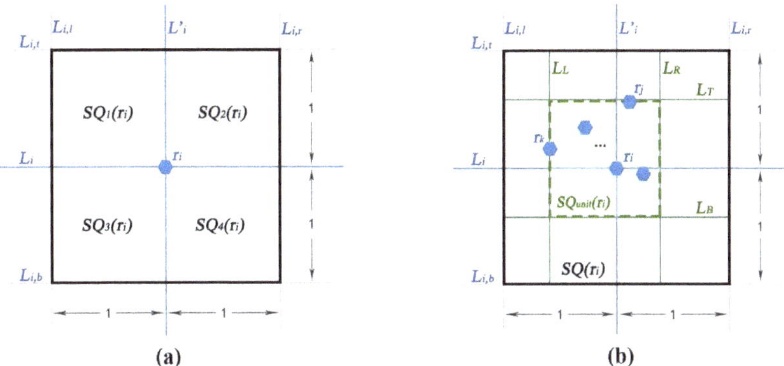

Figure 2. An illustration of (**a**) Square Area and (**b**) Unit Area.

Unit Area. Let r_j, r_k, respectively, be the topmost and leftmost robots among the robots in $SQ(r_i)$. In some situations, both r_j and r_k may be the same robot, and this definition is still valid. Let L_T be the horizontal line passing through r_j and L_L be the vertical line passing through r_k. Let L_B be the horizontal line parallel to L_T, and it is at distance one south of L_T. Similarly, let L_R be the vertical line parallel to L_L and at a distance of one east of L_L. The *unit area* for r_i, denoted as $SQ_{unit}(r_i)$, is an area of the plane inside $SQ(r_i)$ enclosed by lines L_L, L_T, L_R, and L_B. Note that $SQ_{unit}(r_i)$ is an (axis-aligned) unit square of both height and width one. We denote by p_{TL}, p_{BL}, p_{BR}, and p_{TR} the intersection points of lines L_T and L_L, L_B and L_L, L_B and L_R, and L_T and L_R, respectively. Figure 2b illustrates the idea of unit area computation.

Visibility Graph and Gathering Configuration. We define the visibility graph of any initial configuration I and gathering configurations as follows.

Definition 1 (Initial Visibility Graph). *The visibility graph $G(I) = (Q, E)$ of any arbitrary initial configuration I of robots is the graph such that, for any two distinct robots r_i and r_j, $(r_i, r_j) \in E$ where r_j is positioned on or inside $SQ(r_i)$ (or vice-versa).*

$SQ(*)$ provides connectivity for robots with square connectivity range $\sqrt{2}$. The gathering problem may not be solvable under limited visibility if the initial visibility graph $G(I)$ is not connected [2,6]. Therefore, we assume that $G(I)$ is connected at time $t = 0$. Moreover, any algorithm for gathering must maintain the connectivity of $G(I)$ during its execution until a gathering configuration is reached. For the sake of clarity, we denote by $G_t(I)$ the visibility graph $G(I)$ for any time $t \geq 0$.

Definition 2 (Ideal Gathering Configuration). *An ideal gathering configuration is one where all robots are at a single point that is not known beforehand.*

Definition 3 (Relaxed Gathering Configuration). *A relaxed gathering configuration is one where all robots are on a horizontal segment of length 1 unit that is not known beforehand.*

The relaxed gathering configuration (Definition 3) is inspired from the recent work of Cord-Landwehr et al. [14], where the authors modified the ideal gathering configuration (Definition 2) to solve gathering on a grid by locating all robots within a 2×2-sized square area that is not known beforehand. Additionally, Definition 3 helps us to circumvent the impossibility results relative to gathering to a point in the $\mathcal{ASYNC}$ setting [3], even when $N = 2$, by gathering the robots in a unit horizontal line segment. As an example, consider two robots r_i, r_j at distance 1 apart on a horizontal line working under an $\mathcal{ASYNC}$ setting. Let r_i and r_j activate at the same time and r_i moves to the position of r_j and r_j moves to the position of r_i as both of them move in the horizontal line. This scenario may repeat infinitely since r_i and r_j do not have common agreement on east or west under one-axis agreement on north. By using our square connectivity range $\sqrt{2}$, the viewing range $\sqrt{10}$ and one-axis agreement, even when $N = 2$, the robots can reach a horizontal segment of length one unit. The viewing range helps each robot r_i to see whether there is a robot outside $SQ(r_i)$ and decide whether Definition 3 is reached.

Under both axis agreement, our algorithm provides an ideal gathering configuration (Definition 2). Under one-axis agreement, our algorithm provides a relaxed gathering configuration (Definition 3). Since we focus on runtime, we do not explicitly characterize the configurations that do not achieve Definition 2 under one-axis agreement, but we simply prove that all the configurations (at least) attain Definition 3 in $\mathcal{O}(D_E)$ time.

3. $\mathcal{O}(D_E)$ Time Algorithm for the Grid

In this section, we define the grid model that is a restriction imposed on the Euclidean plane. The motivation behind designing an algorithm for this model is that it is simple to understand and easy to analyze. We design and analyze an algorithm without the grid restriction in Section 4. In the grid model, a robot moves on a two-dimensional grid and changes its position to one of its eight horizontal, vertical, or diagonal neighboring grid points. Throughout this section, we assume that robots agree on both coordinate axes, and each robot has the viewing range of two. Moreover, each robot has the square connectivity range of $\sqrt{2}$. We say gathering is performed when the robot configuration satisfies Definition 2.

3.1. The Algorithm

The pseudocode of the algorithm is given in Algorithm 1. Depending on the positions of other robots within its viewing range, r_i distinguishes *diagonal*, *horizontal*, and *vertical hops*, which we discuss separately below. A robot r_i hops on one of its neighboring grid points based on the diagonal, horizontal, or vertical pattern that matches the snapshot it takes in the *Look* phase. Notice that since robots agree on north, r_i never hops on any of the three neighboring grid points relative to north from its position, i.e., r_i hops only to one of its five neighboring grid points on the same horizontal line L_i or south of L_i. We will show that this allows achieving a gathering progress in every epoch. Since robot moves are not instantaneous due to the $\mathcal{ASYNC}$ setting, a robot r_i also does not move if it observes at least one robot in the north of L_i inside or on $SQ(r_i)$. This is crucial for guaranteeing that robots do not move south forever. Robot r_i terminates when it sees no other robot inside or on $SQ(r_i)$ other than its position.

Diagonal Hops. A diagonal hop takes a robot r_i to one of the two diagonal neighboring grid points in the south (i.e., either p_{br} or p_{bl}). Let L_i be a horizontal line that passes from the current position of a robot r_i. Robot r_i makes a diagonal hop when it sees no robot in $SQ(r_i)$ at the north of L_i (including the positions of L_i) and either (i) r_i sees no other robot in $SQ_3(r_i)$ (except at its position) and sees at least one robot on $L_{i,r}$ at the south of L_i or (ii) r_i sees no other robot in $SQ_4(r_i)$ (except at its position) and sees at least one robot on

$L_{i,l}$ at the south of L_i. In case (i), r_i hops on grid point p_{br}, whereas in case (ii), it hops on grid point p_{bl}.

In this hop, the robot moves diagonally at a distance of $\sqrt{2}$. Figure 3a,b illustrate diagonal hops.

Algorithm 1: The algorithm for gathering on a grid (under both axis agreement)

```
   /* In every LCM cycle, each robot r_i does the following when it
      activates:                                                   */
   /* Look:                                                        */
```
1   $(x_i, y_i) \leftarrow$ current position of robot r_i in the grid graph G;
2   $C(r_i) \leftarrow$ snapshot of the positions of other robots within the viewing range of r_i;
```
   /* Compute:                                                     */
```
3   $SQ(r_i) \leftarrow$ square area for robot r_i;
4   $L_i, L_i' \leftarrow$ horizontal and vertical lines passing through r_i, respectively;
5   $L_{i,t}, L_{i,b} \leftarrow$ horizontal lines parallel to L_i and passing through $(x_i, y_i + 1)$ and $(x_i, y_i - 1)$, respectively;
6   $L_{i,r}, L_{i,l} \leftarrow$ vertical lines parallel to L_i' and passing through $(x_i + 1, y_i)$ and $(x_i - 1, y_i)$, respectively;
7   $d_i \leftarrow$ destination point for r_i to move;
8   **If** r_i sees no other robot in any of the neighboring grid points on $SQ(r_i)$ **then**
9   r_i terminates;
10   **Else if** r_i sees at least a robot in $SQ(r_i)$ in North of L_i **then**
11   r_i keeps waiting; $d_i \leftarrow (x_i, y_i)$;   `// do nothing`
```
   /* Check the following two conditions for a diagonal hop.       */
```
12   **Else if** r_i sees no robot in $SQ(r_i)$ on or West of L_i' (except at its position), and sees at least a robot on $L_{i,r}$ that is part of $SQ(r_i)$ in South of L_i **then**   `// Figure 3a`
13     $d_i \leftarrow (x_i + 1, y_i - 1)$;
14   **Else if** r_i sees no robot in $SQ(r_i)$ on or East of L_i' (except at its position), and sees at least a robot on $L_{i,l}$ that is part of $SQ(r_i)$ in South of L_i **then**   `// Figure 3b`
15     $d_i \leftarrow (x_i - 1, y_i - 1)$;
```
   /* Check the following condition for a horizontal hop.          */
```
16   **Else if** r_i sees at least a robot on $(x_i + 1, y_i)$ and sees no other robot in $SQ(r_i)$, except on L_i in the East **then**   `// Figure 3c`
17     $d_i \leftarrow (x_i + 1, y_i)$;
18   **Else**   `// Check either of the following two conditions for a vertical` hop.
19   **If** r_i sees no robot in $SQ(r_i)$ in North of L_i and sees at least a robot r_j on L_i' in South in $SQ(r_i)$ **then**   `// Figure 3d`
20        $d_i \leftarrow (x_i, y_i - 1)$;
21   **Else if** r_i sees no robot in $SQ(r_i)$ in North of L_i and sees at least one robot each on two lines $L_{i,l}$ and $L_{i,r}$ on or South of L_i in $SQ(r_i)$ **then**   `// Figure 3e`
22        $d_i \leftarrow (x_i, y_i - 1)$;
```
   /* Move:                                                        */
```
23   r_i moves to d_i;
```
   /* Note:  Each robot reaches a grid point after the completion of a
      cycle.  But a robot may not necessarily see other robot(s) (which
      is/are moving) only at grid points since the robots may perform
      their LCM cycles at arbitrary times due to the ASYNC setting.
                                                                   */
```

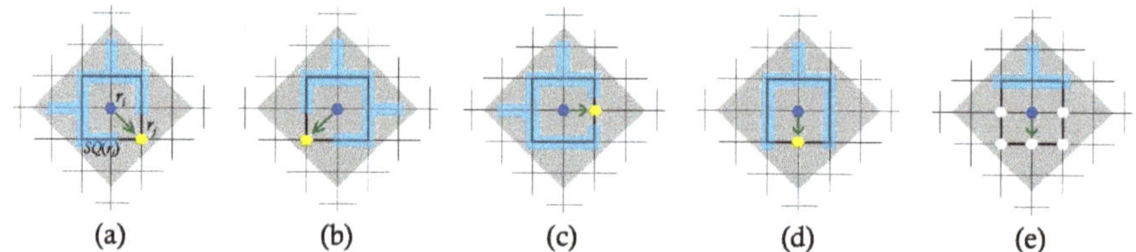

(a)   (b)   (c)   (d)   (e)

Figure 3. An illustration of moves made by a robot: (**a**,**b**) diagonal hops, (**c**) horizontal hop, and (**d**,**e**) vertical hops. The blue shaded area along the grid lines represents that there is no robot in that area. The outer diamond represents the set of grid points within the viewing range of r_i in grid (i.e., 2).

Horizontal Hops. A horizontal hop takes r_i to its neighboring grid point on L_i in the east. When a robot r_i sees at least one robot r_j at its horizontal neighboring grid point (and possibly others on L_i between r_i and r_j) and no other robot in $SQ(r_i)$, r_i makes a horizontal hop to the neighboring grid point in the east. Figure 3c illustrates the horizontal hop.

Vertical Hops. A vertical hop always takes r_i to its neighboring grid point vertically south from it. Robot r_i makes a vertical hop if either (i) it sees a robot r_j on L_i' at the south of L_i and no other robot in $SQ(r_i)$ at the north of L_i or (ii) it sees at least one robot each on $L_{i,l}$ and $L_{i,r}$ or south of L_i and no other robot in $SQ(r_i)$ at the north of L_i. Figure 3d illustrates case (i) and Figure 3e illustrates case (ii).

3.2. Analysis of the Algorithm

We first prove the correctness of the algorithm in the sense that the visibility graph $G_t(I)$ remains connected during execution. We then prove the progress of the algorithm such that after a finite number of epochs, any connected initial configuration converges to an ideal gathering configuration (Definition 2). Let I be any arbitrary initial configuration of robots in Q on a grid such that $G_0(I)$ is connected. Let $SER(I)$ be the *axis-aligned smallest enclosing rectangle* for the robots in I. Let D_Y and D_X, respectively, be the height and width of $SER(I)$. Let $L_{D_Y}, L_{D_Y-1}, \dots, L_0$ be the horizontal line segments of $SER(I)$ at every 1 unit vertical distance, with L_{D_Y} being the topmost horizontal line segment and L_0 being the bottommost horizontal line segment. Similarly, let $L'_{D_X}, L'_{D_X-1}, \dots, L'_0$ be the vertical line segments of $SER(I)$ at every one unit horizontal distance, with L'_{D_X} being the rightmost vertical line segment and L'_0 being the leftmost vertical line segment. Let L_S be the line parallel to L_0 at distance $\frac{D_X}{2}$ south of L_0. Figure 4 illustrates these definitions.

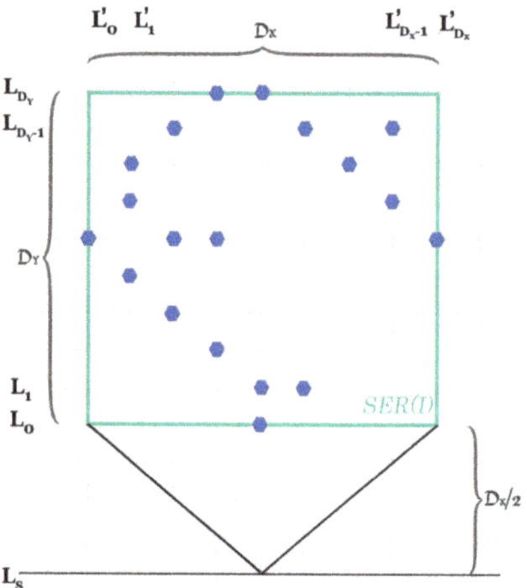

Figure 4. An illustration of an axis-aligned smallest enclosing rectangle $SER(I)$ and the triangular area south of it.

Lemma 1. *Given any initial configuration I such that the visibility graph $G_0(I)$ is connected, the graph $G_t(I)$ at any time $t > 0$ remains connected.*

Proof. For a robot r_i, since $G_0(I)$ is connected, there are robots in at least two out of its eight neighboring grid points, unless r_i is a leaf node in $G_0(I)$ in which case there may be a robot in only one out of its eight neighboring grid points. We will show that no matter the moves of r_i and the robots in its eight neighboring grid points, in the new configuration, r_i has robots in at least two out of its eight neighboring grid points (unless it is a leaf node in $G_0(I)$ in which case there will be a robot in one of eight neighboring grid points). This immediately proves this lemma from our definition of connectivity.

Notice that the movements of r_i are either diagonal, horizontal, or vertical, and r_i never moves to its three neighboring grid points on $SQ(r_i)$ in the north of L_i. Furthermore, r_i does not move when it sees at least one robot r_j at the north of L_i or inside $SQ(r_i)$.

A diagonal hop for r_i is possible when r_i sees other robot(s) r_j only on one of its two diagonal neighboring grid points on $SQ_3(r_i)$ or $SQ_4(r_i)$, and r_i moves to the position of r_j (since r_j does not move as r_j sees r_i at the north of L_j until r_i reaches the position of r_j). Robot r_i also makes a diagonal hop when it sees other robot(s) r_j on either $L_{i,l}$ only or $L_{i,r}$ only that is part of $SQ(r_i)$ in the south of L_i. Since r_j is at the south of L_i, it must be in transit to the neighboring diagonal grid point of r_i and r_i, and r_j meet together when both of them reach that grid point. If one reaches that grid point before, it waits for the other since there will be at least one robot at the north of the horizontal line for the robot on that grid point until r_i and r_j meet.

A horizontal hop for r_i is possible only when r_i sees r_j in the east at the horizontal neighboring grid point (and no other robot inside or on $SQ(r_i)$ except on L_i in the east). After the movement, r_i either reaches the position of r_j (if r_j does not move) or r_i and r_j will be at the two vertical neighboring grid points (if r_j moves). That is, for r_j to move, r_j has to see at least one other robot in addition to r_i on L_i or at the south. The connectivity is maintained since r_i is the endpoint robot (i.e., it has only one neighboring robot), and if r_j is also the endpoint robot, then there is no third robot in the system; otherwise, r_j must see a robot $r_k \neq r_i$ in one of its five neighboring grid points on L_i or south of L_i.

A vertical hop for r_i is possible when it sees at least one other robot r_j on the neighboring grid point that is vertically south of it (and possibly others between r_i and r_j), but no robot is observed on or inside $SQ(r_i)$ in the north of L_i. In this case, r_i reaches the position of r_j since r_j cannot move until there is r_i in the north. r_i performs a vertical hop also when it sees at least one robot each on the two vertical lines $L_{i,l}$ and $L_{i,r}$ on or south of L_i in $SQ(r_i)$ and no robot in $SQ(r_i)$ in the north of L_i. Suppose r_i sees two robots r_j and r_k in $SQ(r_i)$ on or south of L_i such that $r_j \in L_{i,l}$ and $r_k \in L_{i,r}$. After the movement, in this case, the distance between r_i and one of r_j, r_k is at most $\sqrt{2}$ as they will be (at most) at the diagonal neighboring grid points from each other. The lemma is described as follows. $\square$

Lemma 2. *Given any initial configuration I, if all the robots are not at one or two neighboring grid positions on the same horizontal line, the robots on the line segment L_{D_Y} of $SER(I)$ move to the line segment L_{D_Y-1} in at most two epochs.*

Proof. Since L_{D_Y} is the topmost horizontal line segment, there is no robot in the north of L_{D_Y}. Moreover, since robots agree on north, the robots at the grid points of L_{D_Y} never move north of L_{D_Y}. Therefore, if all the robots at the grid points of L_{D_Y} make diagonal or vertical hops when they become active, then they will reach the positions of L_{D_Y-1}; hence, in at most one epoch, all robots on L_{D_Y} will be at L_{D_Y-1}, even in the $\mathcal{ASYNC}$ setting. Note that in an epoch, each robot completes its LCM cycle at least once. This means that, in this case, each activated robot at L_{D_Y} completes its LCM cycle after moving to the position of L_{D_Y-1}. Therefore, a robot r_i on L_{D_Y} remains at a grid point on L_{D_Y} if and only if it makes a horizontal move in that epoch. We will show that r_i either terminates or moves to a position on L_{D_Y-1} in the next epoch.

Let r_j be the robot at the horizontal neighboring grid point that r_i sees on L_{D_Y} when it makes a horizontal hop. We have it that r_i must not have seen any robot on its other seven neighboring grid points or inside of $SQ(r_i)$ (except between r_i and r_j on the same horizontal line). When r_i moves to the position of r_j, either r_j is still on L_{D_Y} or has moved to L_{D_Y-1} in the neighboring grid point that is vertically south of r_j. If r_j has not moved south, either the execution is still in the first epoch or r_j does not see any other robot except on or between the positions of r_i and r_j in the same horizontal line. If r_j is still in the first epoch, then r_i reaches the position of r_j, and either this horizontal moving scenario repeats with execution still being in the first epoch or r_j moves south. If r_j does not see any other robot except on or between the positions of r_i and r_j in the same horizontal line, r_i (and all other robots on L_i up to r_j) reaches the position of r_j, and all of them terminate by achieving the gathering configuration. If r_j has moved to L_{D_Y-1} in the first epoch, r_i moves to the position of r_j on L_{D_Y-1} when it becomes active next time since r_j is in the neighboring grid point of $SQ(r_i)$ that is vertically south of it. Therefore, in at most two epochs, all the robots on L_{D_Y} move to the positions of L_{D_Y-1}. $\square$

The following observation is immediate for vertical hops since a vertical hop by a robot takes it to its neighboring grid point vertically south of it. For a horizontal/diagonal hop, this is also true since a robot performing a horizontal/diagonal hop never finds its neighboring robot outside L'_{D_X} and L'_0.

Observation 1. *No robot of $SER(I)$ moves to the positions outside of lines L'_0 and L'_{D_X} during the execution.*

Lemma 3. *No robot of $SER(I)$ reaches the south of horizontal line L_S (Figure 4) during the execution.*

Proof. Let $\mathcal{X} := \{r_0, \ldots, r_X\}$ be the robots on L_0 in the increasing order of their x-coordinates (some of the grid points on L_0 may be empty, and it does not impact our analysis). If all robots $r_0, \ldots, r_X$ on set $\mathcal{X}$ have robots on or inside $SQ(*)$ at the north of L_0, they do not move until those robots at the north of L_0 are moved to L_0. Therefore, we first assume that only r_0 and r_X have robots at the north of L_0 inside or on $SQ(r_0)$ and $SQ(r_X)$, respectively, and $\{r_1, \ldots, r_{X-1}\}$ have no such robots at the north of L_0 inside or on their respective $SQ(*)$. Robots $r_2, \ldots, r_{X-2}$ can move to their neighboring grid points that are vertically south of them in one epoch. This is because they do not see any robot at the north of the horizontal line passing through their positions.

In the second epoch, r_2 and r_{X-2} see r_1 and r_{X-1}, respectively, in the north on their respective $SQ(*)$, and only the robots $r_3, \ldots, r_{X-3}$ can move to the next line in the south from their current horizontal line. This implies that each robot $r_i, 1 \leq i \leq \frac{X}{2}$ waits for r_{i-1} since it sees r_{i-1} on the neighboring grid point in the north from their position, and this is also the case for the robots from $r_{\frac{X}{2}+1}$ (from $r_{\frac{X}{2}+2}$ in the even D_X case) to r_{X-1}. The scenario repeats until the middle robot of L_0 reaches at most $\frac{D_X}{2} - 1$ distance south from L_0, if D_X is an odd number). If D_X is an even number, two robots $r_{\frac{X}{2}}$ and $r_{\frac{X}{2}+1}$ of L_0 reach at most $\frac{D_X}{2} - 2$ distance south from L_0.

Now consider the case where either r_0 or r_X has no robot on the neighboring grid point that is vertically north from it in addition to $r_1, \ldots, r_{X-1}$. Notice that at least one of r_0 or r_X must have a robot at the north to maintain the connectivity of $G_t(I)$. Let r_0 be that robot (the case of r_X is analogous). If r_0 moves first, it moves to the position of r_1 in L_0 performing a horizontal hop. If r_1 moves first, r_0 reaches r_1 at the horizontal line immediately below L_0 performing a diagonal hop. By repeating this, the robots $r_0, \ldots, r_{\frac{X}{2}-1}$ may reach the position of $r_{\frac{X}{2}}$ (the middle robot) at distance $D_{\frac{X}{2}}$ south of L_0 (i.e., L_S). For the remaining robots $r_{\frac{X}{2}}, \ldots, r_X$, each robot $r_{X-i}, 1 \leq i \leq \frac{X}{2}$ sees r_{X-i+1} in the north on respective $SQ(*)$; thus, the middle robot can reach at most $D_{\frac{X}{2}}$ south of L_0. Therefore, this process again ends up at line L_S in the worst-case.

During the execution, the robots at the north of L_0 in $SER(I)$ may visit the robots south of L_0. In that case, the robots at the south of L_0 do not move until they see at least one robot at the north of its position inside $SQ(*)$. If a robot does not see any robots at the north of its position, then it either performs a diagonal hop which never takes it to the south of L_S or it performs a vertical hop. If it performs a vertical hop, it will perform a horizontal hop in the next epoch, and this never takes it south of L_S. Therefore, according to the moves of the robots of L_0, it is easy to see that all the robots in Q are within the triangular area (as depicted in Figure 4). $\square$

Lemma 4. *The viewings of two and the square connectivity range of $\sqrt{2}$ are sufficient for gathering relative to a grid point (that is not known beforehand) on a grid under both axis agreements.*

Proof. If a robot r_i sees robots only at the south (vertically below or diagonal), it can simply move towards the south, and when r_i sees no robot in the south and no robot on horizontal neighboring grid points, it can simply terminate. This is because if there is another robot within its viewing range, r_i must see it in one of its eight neighboring grid points in order to satisfy connectivity for $G_t(I), t > 0$, (Lemma 1) since $G_0(t)$ satisfies this condition. If r_i sees a robot r_j in either of its horizontal neighboring grid points, then r_i moves to the position of r_j if r_j is at its east, and r_j simply waits for r_i as it does not perform a horizontal hop to the west or moves vertically south. Even in this case, r_i sees r_j. Therefore, if r_i sees no robot in $SQ(r_i)$, it can terminate. According to the definition of the connectivity range, the viewing range of two is enough for r_i to maintain connectivity with any of the eight neighboring grid points. $\square$

The analysis of this section proves the following main result.

Theorem 2. *Given any connected configuration of $N \geq 1$ robots with the viewing range of two and the square connectivity range of $\sqrt{2}$ on a grid, the robots can gather to a point in $\mathcal{O}(D_E)$ epochs in the $\mathcal{ASYNC}$ setting under both axis agreement.*

Proof. We have from Lemma 1 that $G_t(I)$ remains connected during the execution. We have from Lemma 2 that all the robots at the topmost horizontal line L_{D_Y} of $SER(I)$ move to L_{D_Y-1} in at most two epochs. Thus, after at most two epochs, Lemma 2 applies again to the robots of L_{D_Y-1}, which takes all the robots on L_{D_Y-1} to L_{D_Y-2} or south in next two epochs. This process continues and all the robots in $SER(I)$ move to line L_0 or south of it in at most $2 \cdot D_Y$ epochs. These robots will be at one grid point in at most the next D_X epochs. This is because for every one unit of vertical hop of the robots at the south of L_0, the width of the positions of robots decreases by two. The width of the positions of robots at L_0 is at most D_X. Thus, the width of the positions of robots becomes zero at distance $\leq \frac{D_X}{2}$ south of L_0. Again, from Lemma 2, it takes at most two epochs to move all the robots one unit south; hence, to move all the robots at $\frac{D_X}{2}$ distance south of L_0, it takes at most D_X epochs. Therefore, the robots can gather in $\mathcal{O}(D_X + D_Y)$ epochs. We have that $\max\{D_X, D_Y\} \leq D_E \leq \sqrt{2} \cdot \max\{D_X, D_Y\}$ for $SER(I)$ of any initial configuration I. Therefore, $D_X + D_Y \leq 2 \cdot \max\{D_X, D_Y\}$; hence, $\mathcal{O}(D_X + D_Y) = \mathcal{O}(2 \cdot \max\{D_X, D_Y\}) = \mathcal{O}(D_E)$. The algorithm terminates (Lemma 4) since if a robot r_i sees no robot in $SQ(r_i)$ other than its current position, then all the robots of Q must be gathered in the current position of r_i (due to the connectivity guarantee of Lemma 1). $\square$

4. $\mathcal{O}(D_E)$ Time Algorithm for the Euclidean Plane

We discuss here how to solve gathering in a Euclidean plane by removing the restrictions on robot moves imposed on a grid. The viewing range is of $\sqrt{10}$, and the square connectivity range is of $\sqrt{2}$ (both measured in Euclidean distance). The robots agree on both coordinate axes.

We say gathering is performed when the robot configuration satisfies the ideal gathering configuration (Definition 2).

4.1. The Algorithm

The pseudocode of the algorithm is provided in Algorithm 2. Depending on the positions of other robots in its viewing range, a robot r_i can decide to hop on the positions of one of its neighboring quadrants $SQ_3(r_i)$ or $SQ_4(r_i)$; we do not allow r_i to move to the positions north of L_i. In contrast to grid where robots always move in either unit distances (horizontal and vertical hops) or distance $\sqrt{2}$ (diagonal hops), in the Euclidean plane, a robot may move with varying distance of at most one for horizontal and vertical hops and varying distance of at most $\sqrt{2}$ for diagonal hops. The main difference (with the grid) is on how robots match patterns to perform diagonal, horizontal, and vertical hops. In contrast to relatively simple matching patterns of robots on a grid, the matching patterns of robots for the Euclidean plane are complex.

4.1.1. Overview of the Patterns

The idea is to resemble the patterns for the Euclidean plane to the respective patterns for the grid. For this purpose, we ask each robot r_i to compute unit area $SQ_{unit}(r_i)$ as defined in Section 2. $SQ_{unit}(r_i)$ helps r_i to decide whether to make a diagonal, horizontal, or vertical hop. If r_i sees itself or at least one robot in SQ_{unit}, (r_i) is connected to a robot at the north of L_T, and it does not move. This guarantees that robots do not move south forever. If the robots in $SQ_{unit}(r_i)$ are not connected to any other robot outside of $SQ_{unit}(r_i)$ at the west of L_R (or similarly at the east of L_L), then r_i makes a horizontal hop to the east (or similarly to west). If r_i satisfies the conditions for a horizontal hop, except that there is a robot on point p_{BR} (or similarly on p_{BL}), and the robots in $SQ_{unit}(r_i)$ are in a single diagonal line, then it makes a diagonal hop to p_{BR} (or similarly to p_{BL}). If the robots in $SQ_{unit}(r_i)$ are not connected to any other robot outside of $SQ_{unit}(r_i)$ at the north of L_B but (at least) a robot in $SQ_{unit}(r_i)$ is connected to a robot on or south of L_B and r_i does not satisfy a condition for a diagonal hop, then r_i makes a vertical hop. Moreover, if r_i sees at least one robot on each of its two sides (east and west) at horizontal distance ≥ 2, then it makes a vertical hop. The termination is guaranteed by asking r_i to check, in every LCM cycle, whether all robots in its viewing range are positioned in $SQ_{unit}(r_i)$ (that is, r_i sees no robot outside $SQ_{unit}(r_i)$). When that is the case, r_i and the remaining robots in $SQ_{unit}(r_i)$ run a special procedure in order to reach a single point (Definition 2) and terminate their computation. Reaching a single point is facilitated for robots by both axis agreement.

4.1.2. Detailed Description of the Patterns

We provide details of the patterns below. Robot r_i terminates when it sees no other robot in $SQ(r_i)$, except on its current position.

Horizontal Hops. Robot r_i makes a horizontal hop in the following conditions:

- This case is similar to the grid. If r_i sees a robot r_j at its east at distance one on line L_i and there is no robot in $SQ(r_i)$, except the current position of r_i and possibly on L_i from r_i up to r_j, r_i hops to the position of r_j (distance 1).
- Robot r_i hops horizontally east on L_i distance $1 - L_{ik}$ (L_{ik} is the distance between r_i and r_k, the leftmost robot in $SQ(r_i)$) if all the following conditions are satisfied (Figure 5a illustrates this case for a horizontal hop):
  - No robot in $SQ_{unit}(r_i)$ is connected to any other robot at the north of L_T.
  - No robot in $SQ_{unit}(r_i)$ is connected to any other robot at the west of L_R, except for the robots in $SQ_{unit}(r_i)$.
  - There is no robot on L_B of $SQ_{unit}(r_i)$.

Algorithm 2: The algorithm for gathering in the Euclidean plane

---

```
/* In every LCM cycle, each robot r_i does the following when it becomes
   activated:                                                              */
/* Look:                                                                    */
```
1 $(x_i, y_i) \leftarrow$ current position of robot r_i in the plane;
2 $C(r_i) \leftarrow$ snapshot of the positions of other robots within the viewing range of r_i;
```
/* Compute:                                                                 */
```
3 $SQ(r_i) \leftarrow$ square area for robot r_i;
4 $SQ_{unit}(r_i) \leftarrow$ unit area for r_i;
5 $L_i, L_i' \leftarrow$ horizontal and vertical lines passing through r_i, respectively;
6 $L_{i,t}, L_{i,b}, L_{i,r}, L_{i,l} \leftarrow$ top, bottom, right and left boundary lines of $SQ(r_i)$, respectively;
7 $L_T, L_B, L_R, L_L \leftarrow$ top, bottom, right and left boundary lines of $SQ_{unit}(r_i)$, respectively;
8 $d_i \leftarrow$ destination point for r_i to move;
9 **If** r_i sees no robot outside $SQ_{unit}(r_i)$ **then**
10   execute the termination procedure;
11 **If** r_i sees a robot r_j in $SQ_{unit}(r_i)$ that is connected to other robot in North of $L_{i,t}$ (of $SQ(r_i)$) **then**
12   r_i does not move; $d_i \leftarrow (x_i, y_i)$;
```
/* Conditions for horizontal hops                                           */
```
13 **Else if** there is no robot on L_B in the segment of $SQ_{unit}(r_i) \wedge$ no robot in $SQ_{unit}(r_i)$ is connected to any other robot in West of L_R except the robots in $SQ_{unit}(r_i)$ **then**
14   set the destination point d_i as a point at horizontal distance $1 - L_{ij}$ in East (where L_{ij} is the horizontal distance from r_i to the leftmost robot r_j in $SQ_{unit}(r_i)$);
```
       /* Note: If r_i be the leftmost robot in SQ_unit(r_i), then it moves
          distance 1 horizontally in East. And, if the conditions satisfy
          symmetrically, then r_i sets as destination point to the position on
          L_L in West.                                                       */
/* Conditions for diagonal hops                                             */
```
15 **Else if** r_i sees at least a robot on the diagonal point $p_{BR} \wedge$ all the robots in $SQ_{unit}(r_i)$ are in the diagonal line that passes through $SQ_4(r_i) \wedge$ no robot in $SQ_{unit}(r_i)$ is connected to any other robot in the West of L_R, except the robots in $SQ_{unit}(r_i)$ **then**
16   set d_i as the diagonal point p_{BR} at distance $\sqrt{2} - L_{ij}$ (where L_{ij} is the distance from r_i to r_j, the topmost and leftmost robot in $SQ_{unit}(r_i)$);
```
       /* Note: Here, p_BR is the intersection point of L_B and L_R and SQ_4(r_i)
          is the unit square quadrant of SQ(r_i) in the South-West region. If r_i
          be the topmost (leftmost) robot in SQ_unit(r_i), then it moves distance
          √2 diagonally to p_BR. Moreover, if the above conditions satisfy
          symmetrically, then r_i sets as destination point p_BL (the
          intersection point of L_B and L_L).                                */
```
17 **Else** // Conditions for vertical hops
18     $SP_{unit}(r_i) \leftarrow$ unit area in West of $L_{i,l}$ and South of L_B;
19   **If** r_i sees a robot at the intersection point of lines L_i' and $L_B \vee r_i$ sees at least one robot each in both sides (East and West) at horizontal distance $\geq 2 \vee (r_i$ sees a robot on L_B of $SQ_{unit}(r_i)$, no robot in $SQ_{unit}(r_i)$ is connected to other robot in North of L_B and West of $L_L) \vee (r_i$ sees at least one robot in $SQ_{unit}(r_i)$ that is connected to other robot in South of L_B in West of L_R and no robot in $SQ_{unit}(r_i)$ is connected to other robot in North of L_B and West of $L_L) \vee (r_i$ sees at least a robot in $SP_{unit}(r_i)$ and at least a robot in $SQ_{unit}(r_i)$ is connected to a robot in North of L_B and West of $L_L)$ **then**
20   set d_i as the point vertically South at distance $1 - L_{ij}$ on L_B of $SQ_{unit}(r_i)$ (where L_{ij} is the vertical distance from r_i to L_T);
```
          /* Note: If r_i be the topmost robot in SQ_unit(r_i) then it moves
             distance 1 vertically South.                                    */
/* Move:                                                                     */
```
21 r_i moves to d_i;

Since we ask the robots to always move east in a horizontal hop, we do not have a symmetric case for horizontal hops under both axis agreements.

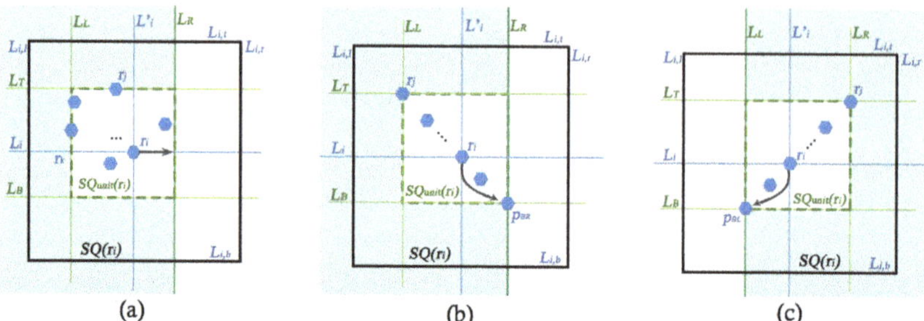

Figure 5. An illustration of a horizontal hop (**a**) and diagonal hops (**b**,**c**).

Diagonal Hops. Robot r_i makes a diagonal hop in either of the following conditions:

- This case is similar to grid. If r_i sees no other robot in $SQ(r_i)$ except at least one robot r_j in $SQ_4(r_i)$ on the diagonal corner point p_{br}, r_i hops to p_{br}. Robot r_i moves at a distance of exactly $\sqrt{2}$ if it performs this hop.
- Robot r_i hops diagonally at a distance of $\sqrt{2} - L_{ij}$ (where L_{ij} is the distance between r_i and r_j, the topmost which is also the leftmost robot of $SQ_{unit}(r_i)$ at point p_{TL}) to a point in $SQ_4(r_i)$, if the following conditions are satisfied:

    – No robot in $SQ_{unit}(r_i)$ is connected to any other robot at the north of L_T.
    – No robot in $SQ_{unit}(r_i)$ is connected to any other robot at the west of L_R, except the robots in $SQ_{unit}(r_i)$.
    – All robots in $SQ_{unit}(r_i)$ are in the diagonal line that passes through $SQ_4(r_i)$.
    – There is at least one robot on the diagonal point p_{BR} of $SQ_{unit}(r_i)$.

Figure 5b illustrates this hop for r_i. The symmetric diagonal case moves r_i to point p_{BL} which is illustrated in Figure 5c.

Vertical Hops. If no robot in $SQ_{unit}(r_i)$ of a robot r_i is connected to any other robot at the north of $L_{i,t}$ (of $SQ(r_i)$), r_i makes a vertical hop of distance $1 - L_{im}$ (where L_{im} is the vertical distance from r_i to line L_T) in either of the following conditions:

- Robot r_i sees at least one robot at the intersection point of L_i' and L_B.
- Robot r_i sees at least one robot each at both the east and west at horizontal distance ≥ 2. Figure 6b illustrates this case.
- Robot r_i sees at least one robot on L_B of $SQ_{unit}(r_i)$, no robot in $SQ_{unit}(r_i)$ is connected to any other robot at the north of L_B and west of L_L, and the conditions for a diagonal hop are not satisfied for r_i. Figure 6a illustrates this case.
- Robot r_i sees at least one robot in $SQ_{unit}(r_i)$ that is connected to a robot at the south of L_B on or west of L_R, and no robot in $SQ_{unit}(r_i)$ is connected to any other robot at the north of L_B and west of L_L. Figure 6a also illustrates this case.
- Let $SP_{unit}(r_i)$ be a unit area at the west of $L_{i,l}$ and south of L_B with L_B being the topmost horizontal line L_T of $SP_{unit}(r_i)$ and $L_{i,l}$ being the rightmost vertical line L_R of $SP_{unit}(r_i)$. Robot r_i sees that at least one robot in $SQ_{unit}(r_i)$ is connected to a robot at the north of L_B and west of L_L, r_i sees at least one robot in $SP_{unit}(r_i)$, and the conditions for a horizontal hop are not satisfied. Figure 6c illustrates this case.

Remark 1. *Robot r_i also makes a vertical hop if the symmetric situations in the last three conditions are satisfied. The above rules infer that the robots move only under certain situations. Robots do not move in all the remaining situations. This process repeats until all robots of Q are inside an (axis-aligned) 1×1-sized square area so that the special procedure for termination, as described in the next paragraph, can be applied.*

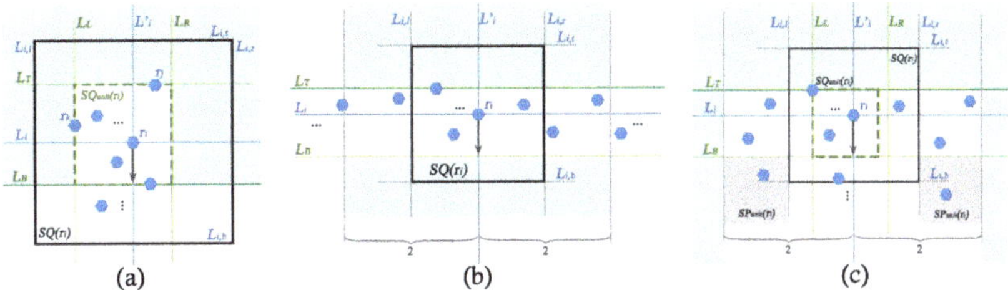

Figure 6. An illustration of vertical hops. (**a**) r_i sees at least one robot on L_B of $SQ_{unit}(r_i)$. (**b**) r_i sees at least one robot each in both sides east and west at horizontal distance ≥ 2. (**c**) r_i does not see any robot at horizontal distance ≥ 2 in both sides east and west, but at least one in the west of L_L (or East of L_R) and is connected to other robot in the south of L_B and west of $L_{i,l}$ (or East of $L_{i,r}$).

4.1.3. The Termination Procedure

We will show in the analysis that the diagonal, horizontal, and vertical hops described above position all robots in Q in an axis-aligned 1×1-sized square area, say SA. We now discuss how the robots reach a point and terminate. Let r_l, r_b, and r_r be the leftmost, bottommost, and rightmost robots in SA. We have that the unit area $SQ_{unit}(r_i)$ of each robot r_i that is in SA overlaps. Therefore, if all the robots in SA are in a single diagonal line, then r_b does not move, and all other robots in SA make a diagonal hop with their destination as the current position of r_b. Otherwise, the robots first perform a horizontal hop, as the destination points the positions on the right vertical line L_R of SA. The robots on L_R do not move until all the robots in SA (the same for all robots) are positioned on L_R. After that, the robots (now on L_R) perform a vertical hop to the destination, which is the position of the bottom most robot on L_R, which does not move. Now, since all the robots reach the same position, they terminate in the next epoch.

We have the following immediate observation after all the robots in Q are positioned in an axis-aligned 1×1-sized square area SA.

Observation 2. *The robots within an axis-aligned 1×1-sized square area SA are positioned at a single point in at most two epochs.*

4.2. Analysis of the Algorithm

We first prove correctness and then progress to providing a guarantee of the algorithm. We use $SER(I)$ and other definitions as in Section 3 except L_S. Here, we define L_S as a horizontal line parallel to L_0 at distance D_X south of L_0. Figure 7 illustrates these definitions for the algorithm in the Euclidean plane.

Based on the movement of robots in horizontal, diagonal, and vertical hops, the following observation is immediately made. This is because the robots never make a horizontal hop to the west, and the robots making the horizontal hops never reach east of L'_{D_X}. In the diagonal hops, robots move to the diagonal position that is closer to the other neighboring robots. Since all the robots are inside L'_0 and L'_{D_X} initially, there is no neighboring robot outside of those lines; hence, the diagonal hops will also be inside L'_0 and L'_{D_X}. In the vertical hops, robots always move vertically south. Since no robot has reached outside of L'_0 and L'_{D_X} with the horizontal as well as diagonal hops, this is true with the vertical hop as well.

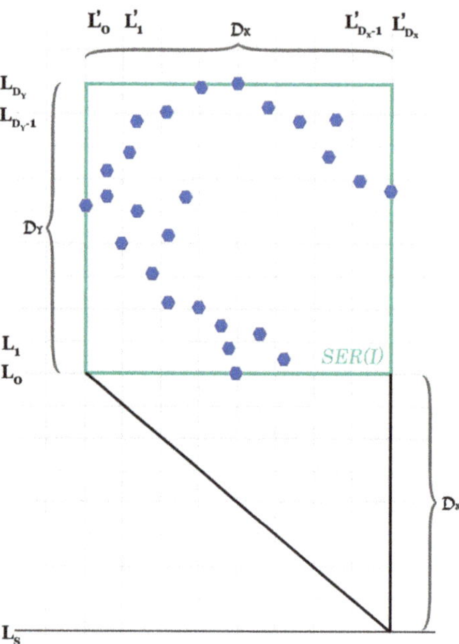

Figure 7. Illustration of $SER(I)$ and the triangular area south of it in the Euclidean plane.

Observation 3. *No robot of $SER(I)$ moves outside of lines L'_0 and L'_{D_X} during the execution.*

Lemma 5. *Given that $G_0(I)$ is connected, the visibility graph $G_t(I)$ at any time $t > 0$ remains connected.*

Proof. We extend the proof of Lemma 1. Similarly to the grid, a robot r_i either does not move or performs either a diagonal, horizontal, or a vertical hop. Note also that r_i never moves to any position at the north of L_i. Furthermore, r_i does not move when it sees at least one robot r_j on line $L_{i,t}$ or at the north of $L_{i,t}$.

First of all, if the robots move as in the grid case, Lemma 1 provides the connectivity proof for $G_t(I), t > 0$, starting from connected $G_0(I)$. Therefore, we focus only on the cases that are particularly relevant to the Euclidean plane.

A diagonal hop for r_i is possible only when the robots in $SQ_{unit}(r_i)$ are in the diagonal line that passes through $SQ_4(r_i)$ and are not connected to any other robot at the west of L_R besides the robots in $SQ_{unit}(r_i)$ (the analogous case of $SQ_3(r_i)$ can be handled similarly). Moreover, there is at least one robot at point p_{BR}. Robot r_i then moves to p_{BR}. This preserves connectivity for $G_t(I)$ since the robot at p_{BR} must be connected to at least one robot at the east of L_R if all the robots of Q are not inside $SQ_{unit}(r_i)$. Due to the $\mathcal{ASYNC}$ setting, the robot r_j at p_{BR} may perform its *Look* phase while r_i is in transit to p_{BR}. Let t' be the time at which r_j performs its *Look*. Let r_i be at point p at distance $\sqrt{2} - x$ from p_{BR} at time t. Let $SQ^p(r_i)$ be $SQ(r_i)$ for r_i when it is at position p. Even in this case, r_j does not move in a position outside of $SQ^p(r_i)$ regardless of whether it performs a horizontal, vertical, or a diagonal hop, preserving connectivity.

A horizontal hop for r_i is possible only when the robots in $SQ_{unit}(r_i)$ are not connected to any other robot at the west of L_R similarly to the diagonal hop case with only the difference being that there is no robot on p_{BR} so that even when all the robots in $SQ_{unit}(r_i)$ are in a diagonal line, r_i cannot perform a diagonal hop. Even in this case, connectivity is preserved since, if the robots move at most the permitted distance, they would have moved in the grid case.

Similarly, in the vertical hop of robot r_i, if it sees no robot at the north in $SQ(r_i)$, it moves vertically south on L'_i with a distance of exactly one. If r_i sees at least one robot r_j at the north in $SQ(r_i)$ and satisfies the conditions for vertical hopping, it hops $1 - L_{ij}$ (where L_{ij} is the vertical distance between r_i and r_j) distance south to a position on L_B (of $SQ_{unit}(r_i)$). In both cases, by the end of first epoch, each robot moves at most a distance of one toward the south, and the robots remain connected because r_i reaches at most $\sqrt{2}$ distance away from the neighbor robot in $SQ(r_i)$ (if the neighbor robot does not move in this epoch) and remains connected. Moreover, the robots connected to r_i at the south of L_B do not move as they find r_i at the north and the connectivity with them remains unaffected. Thus, $G_t(I), t > 0$ remains connected with a vertical hop. Figure 8 illustrates the movement of robots and how the connectivity is preserved. Note here that regardless of the grid case where robots are positioned on the grid points, in the Euclidean plane, robots can be positioned anywhere in the plane. The shaded regions in Figure 8 represent the arbitrary positions of robots within the equivalent grid areas in the Euclidean plane. $\square$

Lemma 6. *All the robots in at the north of L_{D_Y-1} in $SER(I)$ move to the positions on L_{D_Y-1} or south of L_{D_Y-1} in at most three epochs.*

Proof. Since L_{D_Y} is the topmost horizontal line segment of $SER(I)$, there is no robot at the north of L_{D_Y}. Moreover, since robots agree on north, they never move to the north of the horizontal line they are currently positioned. Consider the robots in the corridor area CA of $SER(I)$ formed by horizontal lines L_{D_Y} and L_{D_Y-1}, excluding the positions of L_{D_Y-1}. Note that in the grid case, the robots were either on L_{D_Y} or on L_{D_Y-1}, and we proved in Lemma 2 that the robots on L_{D_Y} reach L_{D_Y-1} or the south in at most two epochs.

Consider $SQ_{unit}(r_i)$ of any robot r_i in CA. We will show that all the robots in $SQ_{unit}(r_i)$ that are in CA reach L_{D_Y-1} or below in at most two epochs. Since these robots do not see any robot on or north of L_{D_Y}, they perform at least one kind of hop (vertical, horizontal, or diagonal) in the first epoch except the robots positioned between L'_1 and L'_2 in the west and L'_{D_X-2} and L'_{D_X-1} in the east in CA (Figure 8). The robots between L'_1 and L'_2 and L'_{D_X-2} and L'_{D_X-1} may not satisfy any conditions for movement in the first epoch. If all the robots in $SQ_{unit}(r_i)$ in CA perform either a diagonal or a vertical hop, they reach L_{D_Y-1} or south in one epoch because, with both the diagonal and vertical hops, robots in $SQ_{unit}(r_i)$ reach L_B or below L_B and L_B is either L_{D_Y-1} or below. If some robots perform a horizontal hop in the first epoch, we show that it performs either a vertical or a diagonal hop in the second epoch.

A robot makes a horizontal hop if it sees no other robot at the west of L_R, except the robots in $SQ_{unit}(r_i)$, and there is no robot on L_B of $SQ_{unit}(r_i)$. By the end of the first epoch, all the robots in $SQ_{unit}(r_i)$ reach the positions on or east of L_R if all of them make a horizontal hop. In this case, the robots in CA between L'_1 and L'_2 do not move in the first epoch. If some robots performed horizontal hops and the rest performed vertical/diagonal hops, we only need to guarantee that the robots that performed a horizontal hop on the first epoch reached L_B or south of it in the second epoch performing a vertical or a diagonal hop. Consider a robot r_i that satisfies the conditions for a horizontal hop in the first epoch. We have it that there is no robot on L_B, and the robots in $SQ_{unit}(r_i)$ are connected to no other robot besides the robots in $SQ_{unit}(r_i)$ at the west of L_R. Let $SQ^E_{unit}(r_i)$ be a square area adjacent to $SQ_{unit}(r_i)$ at the east between lines L_T and L_B. Let L'_B and L'_R be the bottom horizontal and right vertical lines of $SQ^E_{unit}(r_i)$. If the robots in $SQ^E_{unit}(r_i)$ are not on L'_B and not connected to any other robot below L'_B at the west of L'_R, they do not move until all the robots in $SQ_{unit}(r_i)$ reach L_R or east of L_R in $SQ^E_{unit}(r_i)$. If there are robots on L'_B, let x be the robot on L'_B that is the closest from L_R. Let L_V be a line parallel to L_R at some unit distance west of x. All the robots in $SQ_{unit}(r_i)$ at the west of L_V perform one horizontal move each in the first epoch. The robots on L_V or east in $SQ_{unit}(r_i)$ perform a vertical hop as there are robots on L'_B. The robots of $SQ_{unit}(r_i)$ that performed a horizontal hop on the first epoch now observe robots on L_B in the second epoch and make a vertical move. Moreover, the robots in $SQ^E_{unit}(r_i)$, which were waiting for the robots in $SQ_{unit}(r_i)$ to perform a horizontal hop in the first epoch, now see robots at their respective L_B and, hence, perform either a vertical or a diagonal hop to L_B. This means that the robots in CA between L'_0 and L'_2 also move to L_{D_Y-1} or south in two epochs. Arguing similarly, if the robots in CA between L'_{D_X-2} and L'_{D_X-1} do not move in the first epoch, they also see a robot on their respective L_B in the second epoch since the robots at the west of L'_{D_X-2} have already moved south in the first epoch. Thus, these robots also move south in the second epoch. In the same manner the remaining robots between L'_{D_X-1} and L'_{D_X} in CA move to L_{D_Y-1} or move south in the third epoch as they can observe at least one robot on their respective L_B. Therefore, in three epochs, all the robots in CA reach L_{D_Y-1} or south. The Lemma is described follows. $\square$

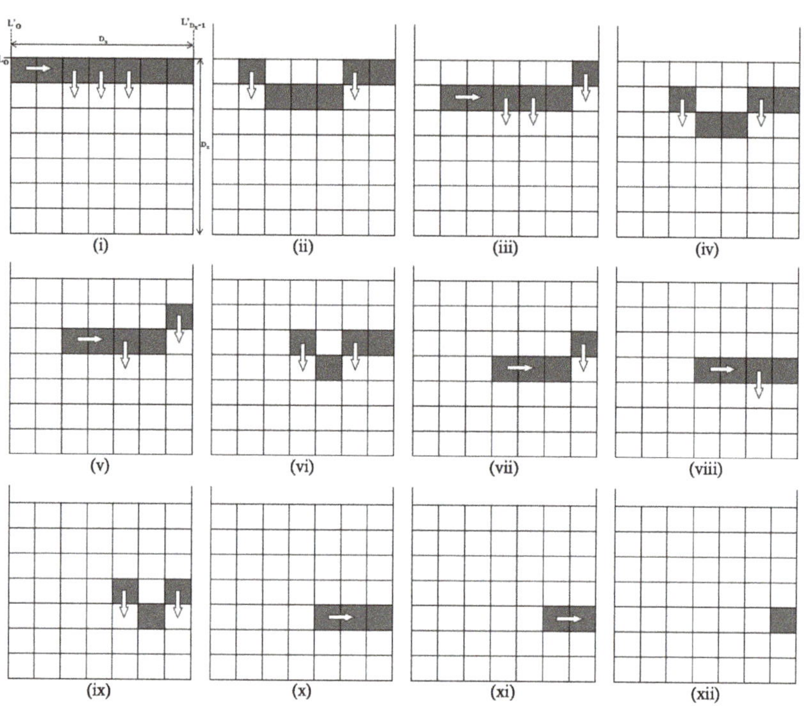

Figure 8. An illustration of movements of robots in the Euclidean plane below L_0. The horizontal and vertical lines are separated by 1 unit distance away, and the robots are positioned arbitrarily in the shaded regions (i.e., they do not need to be necessarily always on the horizontal or vertical lines as in the grid case). At every one unit south of L_0, the width of the positions of robots decreases by 1 unit; hence, all the robots reach inside a unit square at most D_X unit south of L_0. (**i**) All the robots reached South of L_0. (**ii**)–(**xi**) Movements of robots in the South of L_0 in each round. (**xii**) Robots gathered inside a unit square area in the South of L_0.

Lemma 7. *No robots of $SER(I)$ reaches south of L_S during the execution.*

Proof. We extend the proof of Lemma 3. Let $\mathcal{X} := \{r_0, \ldots, r_X\}$ be the set of robots in the increasing order of their x-coordinates in the corridor area CA between L_1 and L_0 of $SER(I)$. If the robots on set $\mathcal{X}$ see other robots at distance ≥ 1 north from their positions, they do not move and wait until they do not see any robot at the north at a distance ≥ 1. Therefore, similarly to Lemma 3, the robots in CA that do not see any robot in the north at distance ≥ 1 proceed to move south. This will take those robots to the next corridor CA' adjacent to CA in the south. Suppose at some time $t > 0$, all the robots reach CA

between L_0 and L_1. Some robots might see no robots at the north at the distance ≥ 1 before time t, and they can perform their moves earlier, which does not affect our argument. Now, the robots in CA that are in a unit square area (i.e., between L_0' and L_1') in the east perform horizontal moves, and the robots in the next unit square area (i.e., between L_1' and L_2') do not move in the first epoch. Similarly, the robots in CA that are in two unit square areas in the west (i.e., between L_{D_X-2}' and L_{D_X}') do not move in the first epoch. The remaining robots in CA between L_2' and L_{D_X-2}' move south. In the next epoch, the robots in CA between L_1' and L_2' (including the robots that moved horizontally to this area in the previous epoch) move south. The robots in CA between L_{D_X-2}' and L_{D_X-1}' also move south in the this epoch whereas the robots in CA between L_{D_X-1}' and L_{D_X}' still do not move; they move in the third epoch. The robots in CA' between L_2 and L_4 and L_{D_X-4} and L_{D_X-2} do not move first, and the other robots move south.

Following this, we can observe that as the robots move to the next corridor (of size one) in the south of L_0, the width of the positions of robots decreases by one. This is because, at every corridor, the robots in the east most unit square area perform horizontal moves. Therefore, all the robots in $\mathcal{X}$ will be within a single unit square area in the corridor at distance D_X south of L_0 (at most). When all the robots are within a unit square area, they follow the termination procedure and do not move further south.

Figure 8 illustrates how the robots move south of L_0. The figure also shows how the robot chains merge to eventually reach a unit square during execution so that the termination procedure can be executed. $\square$

The following observation is also immediate.

Observation 4. *For every one unit vertical hop of the robots in Q in the south of L_0, the width of the positions of robots decreases by (at least) one.*

Lemma 8. *The viewing range of $\sqrt{10}$ is sufficient for gathering to a point (that is not known beforehand) on a plane under both axis agreements.*

Proof. Let r be a robot in Q. $SQ_{unit}(r)$ is computed based on the position of other robots in $SQ(r)$, which may lie anywhere within $SQ(r)$. For r to decide whether it is connected to other robots outside $SQ_{unit}(r)$, it has to see other robots in both the horizontal and vertical distance of at most one outside $SQ_{unit}(r)$. Therefore, the maximum distance between r and some other robot r' in $SQ_{unit}(r)$ (or $SQ(r)$) is $\sqrt{2}$ and r' may be connected to a robot at a distance of at most $\sqrt{2}$ away from r'. Therefore, r needs to see at most a distance of $\sqrt{2} + \sqrt{2} = 2\sqrt{2} = \sqrt{8}$ to find out whether there is a robot outside $SQ_{unit}(r)$ or not. When r sees that no robot in $SQ_{unit}(r)$ is connected outside of $SQ_{unit}(r)$, it can execute the termination procedure.

Now, for the vertical hops, there is one condition that requires r to see at least one robot each at horizontal distance ≥ 2 at both the east and west, within the corridor of L_T and L_B. To guarantee whether there is a robot at horizontal distance ≥ 2 or not, r needs to see up to a horizontal distance of < 3 and vertical distance of < 1. This is because if there is any robot at horizontal distance > 2, it must be connected to a robot at horizontal distance < 2. Therefore, r needs to see at most distance $\sqrt{3^2 + 1^2} = \sqrt{10}$. Figure 9 (left) illustrates this requirement. $\square$

The analysis of this section proves the following main result.

Theorem 3. *Given any connected configuration of $N \geq 1$ robots with the viewing range of $\sqrt{10}$ and the square connectivity range of $\sqrt{2}$ on a plane, the robots can gather to a point in $\mathcal{O}(D_E)$ epochs in the $\mathcal{ASYNC}$ setting under both axis agreements.*

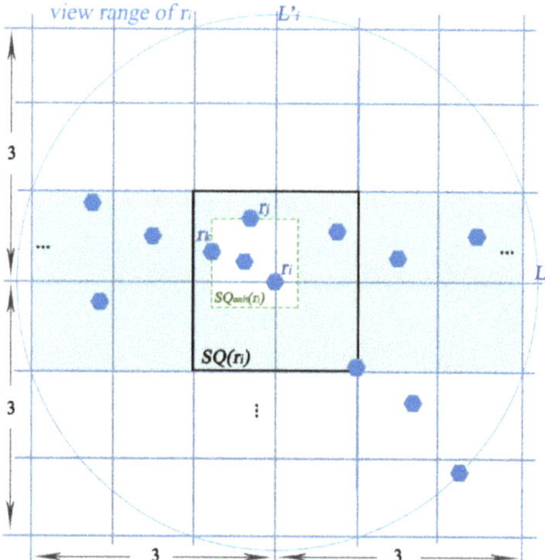

Figure 9. An illustration of the viewing range of $\sqrt{10}$.

Proof. We have from Lemma 5 that, given a connected $G_0(I)$, $G_t(I)$, $t > 0$, remains connected during the execution of the algorithm. We have from Lemma 6 that all the robots at the topmost horizontal line L_{D_Y} of $SER(I)$ move to L_{D_Y-1} or south of L_{D_Y-1} in at most two epochs. In other words, L_{D_Y-1} becomes L_{D_Y} in at most two epochs and Lemma 6 applies again to L_{D_Y-1}. Therefore, all the robots in $SER(I)$ move to line L_0 or south of it in at most $2 \cdot D_Y$ epochs. After that, we have it that, from Lemma 7, these robots will be inside an axis-aligned unit area in at most next $2 \cdot D_X$ epochs, arguing similarly with respect to Lemma 6. After all the robots of Q reach the insides of an unit square area, we have from Observation 2 that they reach a single point in at most the next two epochs. Therefore, the robots gather to a single point in $2 \cdot D_Y + 2 \cdot D_X + 2 = \mathcal{O}(D_X + D_Y) = \mathcal{O}(D_E)$ epochs. The algorithm terminates (Lemma 8). $\square$

5. Gathering under One-Axis Agreement

We discuss modifying the above algorithms when the robots agree on only one axis.

5.1. Grid

We first discuss changes in the model of Section 3. We say gathering is performed when the robot configuration satisfies the relaxed gathering configuration (Definition 3). We also relax the viewing range from 2 to 3.

We now discuss changes in Algorithm 1 (Section 3). The change is only on Rules 1 (termination) and 3 (horizontal hop). Regarding Rule 3, instead of r_i moving only to the east (Figure 3 (middle)), r_i can also move to the west as well if it sees no robots on or inside $SQ(r_i)$, except for the situation where there is exactly one robot r_j on the neighboring grid point on L_i in the west. Regarding Rule 1, r_i terminates if it sees all the robots at at most one unit apart in a horizontal line (i.e., all the robots are positioned in two horizontal neighboring grid points).

Lemma 9. *The viewing range of three is sufficient for gathering on a grid with guaranteed termination under a one-axis agreement.*

Proof. Notice that a robot r_i terminates if it sees all the robots in Q are at most two neighboring grid points (one is current position of r_i and the other is either the left horizontal grid point only or the right horizontal neighboring grid point only). For r_i to make a decision that the robots are not at any third grid point, it has to see all the neighboring grid points of its two horizontal neighboring grid points as well. The distance from r_i to either of its horizontal neighboring grid point is one, and the distance of the neighboring grid points of r_i's horizontal neighboring grid points is at most $\sqrt{2}$. Therefore, r_i needs the viewing range of $(1 + \sqrt{2}) < 3$. The connectivity range remains $\sqrt{2}$. $\square$

Having the viewing range of three, the analysis of the algorithm in Section 3 applies directly to the modified algorithm for the grid under the one axis agreement. Therefore, we summarize the main result in the following theorem.

Theorem 4. *Given any connected configuration of $N \geq 1$ robots with the viewing range of three and the square connectivity range of $\sqrt{2}$ on a grid, the robots can gather in a unit length horizontal line segment (that is not known beforehand) in $\mathcal{O}(D_E)$ epochs in the $\mathcal{ASYNC}$ setting under a one-axis agreement.*

5.2. Euclidean Plane

We first discuss changes in the model of Section 4. We say gathering is performed when the configuration satisfies the relaxed gathering configuration (Definition 3). The viewing and square connectivity ranges remain the same as in Section 4.

We now discuss changes in the algorithm. The change is on horizontal and vertical hops and on termination. Instead of computing $SQ_{unit}(r_i)$ using L_L and L_T as reference lines, $SQ_{unit}(r_i)$ also needs to be computed by using L_R and L_T as references. When r_i sees no other robot on one side (say west) at a distance of >1 but does on the other side (east), it takes the topmost robot r_j and leftmost robot r_k in $SQ(r_i)$ in order to compute $SQ_{unit}(r_i)$; for the symmetric case, it takes the topmost and rightmost robots in $SQ(r_i)$ as a reference. This allows the robots to make horizontal hops in both directions (not necessarily only east under both axis agreement). Therefore, r_i hops to the west of L_i if the conditions for horizontal hop defined in Section 4 are satisfied symmetrically. Regarding vertical hopping, the following changes are made in the last three conditions:

- Robot r_i sees at least one other robot each on both sides of L_i' on L_B or south of L_B, which is connected to at least one robot of $SQ_{unit}(r_i)$.
- Robot r_i sees at least one other robot on L_B or south of L_B (which is connected $SQ_{unit}(r_i)$) at one side of L_i' (say east) and at least one other robot at horizontal distance ≥ 2 on the other side (west) (and vice-versa).
- Robot r_i sees other robot(s) on L_B (or connected to other robot(s) at the south of L_B) only at one side of L_i', say east, then finds the leftmost robot r_l on L_B of $SQ_{unit}(r_i)$ (or south of L_B that is connected to $SQ_{unit}(r_i)$) and sees that no robot in $SQ_{unit}(r_i)$ is connected to another robot at its left (i.e., west) at a horizontal distance of ≥ 1 from r_l (and vice-versa).

Regarding termination, r_i terminates if all the robots it sees within its viewing range (including itself) are within a horizontal line segment of length 1. We will show in the analysis that, with these changes, the algorithm positions the robots in Q inside an axis-aligned 1×1-sized square area SA in $\mathcal{O}(D_E)$ epochs.

We now discuss how the robots in SA reached a relaxed gathering configuration (Definition 3). Let r_b be the bottommost robot in SA (if more than one, pick one arbitrarily). Let L_B be the horizontal line passing through r_b. The robots on L_B (including r_b) do not move. The other robots move vertically to the positions of L_B. The viewing range allows the robots to decide whether there are robots outside SA or not.

Proof of Theorem 1: It is easy to see from the analysis of Section 4 that robots in Q reach inside an axis-aligned unit square area in $\mathcal{O}(D_E)$ epochs. The only change on the analysis is on horizontal hops, which does not increase the number of epochs for the robots in Q to reach the inside of the unit area. Finally, it takes at most one additional epoch for all the robots that are in the unit square area to reach L_B. The robots that are not on L_B move vertically to L_B, and the robots on L_B do not move. Therefore, the robots reach a relaxed gathering configuration (Definition 3) in $\mathcal{O}(D_E)$ epochs. $\square$

6. Concluding Remarks

We have presented, to the best of our knowledge, the first time-optimal $\mathcal{O}(D_E)$-epoch algorithm for gathering $N \geq 1$ classic oblivious robots in a plane in the $\mathcal{ASYNC}$ setting under limited visibility, improving significantly on the previous $\mathcal{O}(D_G)$-round algorithm of [4] that works in the $\mathcal{FSYNC}$ setting. Our result assumes the viewing range of $\sqrt{10}$, the square connectivity range of $\sqrt{2}$, and the agreement on one axis. This is in contrast to the viewing range of one and the (circular) connectivity range of $1 - \frac{1}{\sqrt{2}}$ in [4] under the same one axis agreement. For future work, it will be interesting to relax our assumption of rigid moves to accommodate non-rigid moves. It will also be interesting to reduce the gap between the connectivity and viewing ranges without affecting time complexity.

Author Contributions: Conceptualization, G.S.; methodology, P.P. and G.S.; formal analysis, P.P. and G.S.; investigation, P.P. and G.S.; resources, G.S.; writing—original draft preparation, P.P.; writing—review and editing, G.S.; supervision, G.S.; project administration, G.S. All authors have read and agreed to the published version of the manuscript.

Funding: This research received no external funding

Acknowledgments: The authors thank Costas Busch for introducing this problem.

References

1. Suzuki, I.; Yamashita, M. Distributed Anonymous Mobile Robots: Formation of Geometric Patterns. *SIAM J. Comput.* **1999**, *28*, 1347–1363. [CrossRef]
2. Flocchini, P.; Prencipe, G.; Santoro, N. Distributed Computing by Oblivious Mobile Robots. *Synth. Lect. Distrib. Comput. Theory* **2012**, *3*, 1–185. [CrossRef]
3. Prencipe, G. Impossibility of Gathering by a Set of Autonomous Mobile Robots. *Theor. Comput. Sci.* **2007**, *384*, 222–231. [CrossRef]
4. Izumi, T.; Kawabata, Y.; Kitamura, N. Toward Time-Optimal Gathering for Limited Visibility Model. 2015. Available online: https://sites.google.com/site/micromacfrance/abstract-tasuke (accessed on 18 October 2021).
5. Cieliebak, M.; Flocchini, P.; Prencipe, G.; Santoro, N. Solving the Robots Gathering Problem. In Proceedings of the 30th International Colloquium on Automata, Languages, and Programming, Eindhoven, The Netherlands, 30 June–4 July 2003; pp. 1181–1196.
6. Flocchini, P.; Prencipe, G.; Santoro, N.; Widmayer, P. Gathering of Asynchronous Robots with Limited Visibility. *Theor. Comput. Sci.* **2005**, *337*, 147–168. [CrossRef]
7. Prencipe, G. Autonomous Mobile Robots: A Distributed Computing Perspective. In Proceedings of the 9th International Symposium on Algorithms and Experiments for Sensor Systems, Wireless Networks and Distributed Robotics, Sophia Antipolis, France, 5–6 September 2013; pp. 6–21.
8. Souissi, S.; Défago, X.; Yamashita, M. Gathering Asynchronous Mobile Robots with Inaccurate Compasses. In Proceedings of 10th on Principles of Distributed Systems, Bordeaux, France, 12–15 December 2006; pp. 333–349. [CrossRef]
9. Cieliebak, M.; Flocchini, P.; Prencipe, G.; Santoro, N. Distributed Computing by Mobile Robots: Gathering. *SIAM J. Comput.* **2012**, *41*, 829–879. [CrossRef]
10. Agathangelou, C.; Georgiou, C.; Mavronicolas, M. A Distributed Algorithm for Gathering Many Fat Mobile Robots in the Plane. In Proceedings of the 2013 ACM Symposium on Principles of Distributed Computing, Montréal, QC, Canada, 22–24 July 2013; pp. 250–259.
11. Degener, B.; Kempkes, B.; Meyer auf der Heide, F. A Local O(n^2) Gathering Algorithm. In Proceedings of the Twenty-Second Annual ACM Symposium on Parallelism in Algorithms and Architectures, Thira, Greece, 13–15 June 2010; pp. 217–223.
12. Degener, B.; Kempkes, B.; Langner, T.; Meyer auf der Heide, F.; Pietrzyk, P.; Wattenhofer, R. A Tight Runtime Bound for Synchronous Gathering of Autonomous Robots with Limited Visibility. In Proceedings of the Twenty-Third Annual ACM Symposium on Parallelism in Algorithms and Architectures, San Jose, CA, USA, 4–6 June 2011; pp. 139–148.
13. Kempkes, B.; Kling, P.; Meyer auf der Heide, F. Optimal and Competitive Runtime Bounds for Continuous, Local Gathering of Mobile Robots. In Proceedings of the Twenty-Fourth Annual ACM Symposium on Parallelism in Algorithms and Architectures, Pittsburgh, PA, USA, 25–27 June 2012; pp. 18–26.
14. Cord-Landwehr, A.; Fischer, M.; Jung, D.; Meyer auf der Heide, F. Asymptotically Optimal Gathering on a Grid. In Proceedings of the 28th ACM Symposium on Parallelism in Algorithms and Architectures, Pacific Grove, CA, USA, 11–13 July 2016; pp. 301–312. [CrossRef]
15. Castenow, J.; Fischer, M.; Harbig, J.; Jung, D.; Meyer auf der Heide, F. Gathering Anonymous, Oblivious Robots on a Grid. *Theor. Comput. Sci.* **2020**, *815*, 289–309. [CrossRef]

16. Poudel, P.; Sharma, G. *Universally Optimal Gathering Under Limited Visibility*; Lecture Notes in Computer Science; Springer: Berlin/Heidelberg, Germany, 2017; Volume 10616, pp. 323–340. [CrossRef]
17. Flocchini, P.; Prencipe, G.; Santoro, N. (Eds.) *Distributed Computing by Mobile Entities, Current Research in Moving and Computing*; Lecture Notes in Computer Science; Springer: Berlin/Heidelberg, Germany, 2019; Volume 11340. [CrossRef]
18. Ando, H.; Suzuki, I.; Yamashita, M. Formation and Agreement Problems for Synchronous Mobile Robots with Limited Visibility. In Proceedings of Tenth International Symposium on Intelligent Control, Monterey, CA, USA, 27–29 August 1995; pp. 453–460. [CrossRef]
19. Kirkpatrick, D.; Kostitsyna, I.; Navarra, A.; Prencipe, G.; Santoro, N. Separating Bounded and Unbounded Asynchrony for Autonomous Robots: Point Convergence with Limited Visibility. In *Proceedings of the 2021 ACM Symposium on Principles of Distributed Computing*; ACM: New York, NY, USA, 2021; pp. 9–19. [CrossRef]
20. Pagli, L.; Prencipe, G.; Viglietta, G. Getting Close Without Touching: Near-gathering for Autonomous Mobile Robots. *Distrib. Comput.* **2015**, *28*, 333–349. [CrossRef]
21. Bhagat, S.; Mukhopadhyaya, K.; Mukhopadhyaya, S. Computation Under Restricted Visibility. In *Distributed Computing by Mobile Entities: Current Research in Moving and Computing*; Springer International Publishing: Cham, Switzerland, 2019; pp. 134–183. [CrossRef]
22. Sharma, G.; Busch, C.; Mukhopadhyay, S.; Malveaux, C. Tight Analysis of a Collisionless Robot Gathering Algorithm. *ACM Trans. Auton. Adapt. Syst.* **2017**, *12*, 3:1–3:20. [CrossRef]
23. Lukovszki, T.; auf der Heide, F.M. Fast Collisionless Pattern Formation by Anonymous, Position-Aware Robots. In Proceedings of the 18th Principles of Distributed Systems, Cortina d'Ampezzo, Italy, 16–19 December 2014; pp. 248–262.
24. Cord-Landwehr, A.; Degener, B.; Fischer, M.; Hüllmann, M.; Kempkes, B.; Klaas, A.; Kling, P.; Kurras, S.; Märtens, M.; Meyer auf der Heide, F.; et al. Collisionless Gathering of Robots with an Extent. In Proceedings of the 37th Conference on Current Trends in Theory and Practice of Computer Science, Nový Smokovec, Slovakia, 22–28 January 2011; pp. 178–189.
25. Braun, M.; Castenow, J.; auf der Heide, F.M. Local Gathering of Mobile Robots in Three Dimensions. In *SIROCCO*; Lecture Notes in Computer Science; Springer: Berlin/Heidelberg, Germany, 2020; Volume 12156, pp. 63–79. [CrossRef]
26. Di Stefano, G.; Navarra, A. Optimal Gathering on Infinite Grids. In Proceedings of the 16th Symposium on Self-Stabilizing Systems, Paderborn, Germany, 28 September–1 October 2014; pp. 211–225.
27. Di Stefano, G.; Navarra, A. Optimal gathering of oblivious robots in anonymous graphs and its application on trees and rings. *Distrib. Comput.* **2017**, *30*, 75–86. [CrossRef]
28. D'Angelo, G.; Stefano, G.D.; Klasing, R.; Navarra, A. Gathering of Robots on Anonymous Grids without Multiplicity Detection. In Proceedings of the 19th International Colloquium on Structural Information and Communication Complexity, Reykjavik, Iceland, 30 June–2 July 2012; pp. 327–338. [CrossRef]
29. Cord-Landwehr, A.; Degener, B.; Fischer, M.; Hüllmann, M.; Kempkes, B.; Klaas, A.; Kling, P.; Kurras, S.; Märtens, M.; Meyer auf der Heide, F.; et al. A New Approach for Analyzing Convergence Algorithms for Mobile Robots. In Proceedings of the 38th International Colloquium on Automata, Languages, and Programming, Zurich, Switzerland, 4–8 July 2011; pp. 650–661. [CrossRef]
30. Cohen, R.; Peleg, D. Convergence Properties of the Gravitational Algorithm in Asynchronous Robot Systems. *SIAM J. Comput.* **2005**, *34*, 1516–1528. [CrossRef]
31. Izumi, T.; Potop-Butucaru, M.G.; Tixeuil, S. Connectivity-preserving Scattering of Mobile Robots with Limited Visibility. In Proceedings of the 12th Symposium on Self-Stabilizing Systems, New York, NY, USA, 20–22 September 2010; pp. 319–331.

Article

Robot Evacuation on a Line Assisted by a Bike

Khaled Jawhar [1] **and Evangelos Kranakis** [1,2,*]

[1] School of Computer Science, Carleton University, Ottawa, ON K1S 5B6, Canada;
KhaledJawhar@cmail.carleton.ca
[2] Natural Sciences and Engineering Research Council of Canada Discovery Grant, Ottawa,
ON K1A 1H5, Canada
* Correspondence: kranakis@scs.carleton.ca

Abstract: Two robots and a bike are initially placed at the origin of an infinite line. The robots are modelled as autonomous mobile agents whose communication capabilities are either in the wireless or face-to-face model, while the bike neither can move nor communicate on its own. Thus, the bike is not autonomous but rather requires one of the robots to ride it. An exit is placed on the line at distance d from the origin; the distance and direction of the exit from the origin is unknown to the robots. Only one robot may ride the bike at a time and the goal is to evacuate from the exit in the minimum time possible as measured by the time it takes the last robot to exit. The robots can maintain a constant walking speed of 1, but when riding the bike they can maintain a constant speed $v > 1$ (same for both robots). We develop algorithms for the evacuation of the two robots from the unknown exit and analyze the evacuation time defined as the time it takes the second robot to evacuate. In the wireless model we present three algorithms: in the first the robots move in opposite direction with max speed, in the second with a specially selected "optimal" speed, and in the third the robot imitates the biker (i.e., robot riding the bike). We also give three algorithms in the Face-to-Face model: in the first algorithm the robot pursues the biker, in the second the robot and the biker use zig-zag algorithms with specially chosen expansion factors, and the third algorithm establishes a sequence of specially constructed meeting points near the exit. In either case, the optimality of these algorithms depends on $v > 1$. We also discuss lower bounds.

Keywords: arrival time; bike; evacuation; line; robots; search; speed; optimal trajectory

Citation: Jawhar, K.; Kranakis, E. Robot Evacuation on a Line Assisted by a Bike. *Information* **2021**, *12*, 28. https://doi.org/10.3390/info12010028

Received: 10 November 2020
Accepted: 5 January 2021
Published: 12 January 2021

1. Introduction

Recent years have witnessed an explosive growth of research studies on search from the perspective of mobile agent computing. One of the reasons is because one finds countless natural applications of search and exploration in distributed systems in order to facilitate information exchange between communicating entities. Moreover, there are also applications in numerous other computing areas such as data mining, web crawlers, monitoring and surveillance, just to mention a few.

Evacuation, which is the main theme of our present investigation, is related to search in that one is also interested in searching and exploring a domain in order to find a target. However evacuation usually involves many cooperating entities forming an ensemble or group all of whose members are searching simultaneously by exchanging information (according to a predefined communication model); and unlike search which typically involves only one agent, it is aiming to optimize the arrival time of the last entity in the ensemble. There are many factors that affect how linear search and evacuation problems are solved. Let us assume that the exit is located on a line at a distance d from the origin where the robot starts. The orientation represents the direction that the robot must proceed to reach the exit. The simplest case would be if a robot moving with unit speed knows the distance and the orientation and can thus reach the exit in time d. In this case, the competitive ratio will be the time needed by the robot to reach the exit, which is d, divided by the time needed by the adversary to head to the exit directly, which is d as well. Thus the competitive ratio will be 1. A more complicated case would be if the distance is known and the orientation is not known. As a worst case scenario, it may take the robot $3d$ to find the exit since the robot may move d in the wrong direction and thus it will need to switch the direction and move back $2d$ to reach the exit. The competitive ratio in this case is 3. The problem is even more complicated if both the distance and the orientation are not known. The robot starts at the origin and can move with speed 1. The robot needs to explore both directions in order to find the exit. The best way to achieve this goal is to select a direction and move distance 1. If the exit is not found, the robot will reverse direction and move double the previous distance up until the exit is found. The movement, which is repeated periodically, and which uses a sequence of positive distances that specifies the turning points, is called the Zig-Zag search algorithm. The competitive ratio for the Zig-Zag search algorithm is known to be 9 [1]. Most of the linear search and evacuation problems in the literature were studied using single or multiple robots. Introducing a tool such as a bike to aid the robots was not considered before in any previous work. The study of this paper is based on a new paradigm concerning two robots (also called hikers) aided by a bike and searching for an unknown exit placed somewhere on an infinite line. More specifically, in the "bike assisted evacuation" problem, the hikers and the bike all start at the origin and want to evacuate from an exit placed at an unknown distance and direction (either left or right from the origin) on the infinite line. Evacuation means that eventually both robots must find the exit by reaching its exact location (not necessarily at the same time) on the infinite line. The quality of an evacuation algorithm is measured by the time it takes the second hiker to find the exit, which is also referred to as evacuation time of the ensemble.

1.1. Model and Notation

To analyze the problem proposed, first we describe details concerning mobility and communication of the hikers and describe the role that the bike will play in improving the overall evacuation time.

Mobility and Trajectories. The infinite line is the search domain. It is bidirectional in that the hikers can move in either direction without this affecting their speeds. The hikers can stop at any time and wait as long as they wish, can walk with maximum speed 1 or may ride the bike with speed $v > 1$. An evacuation algorithm is a complete description of the trajectories traced by the two hikers either waiting, walking or riding the bike until they both find the exit. Throughout this paper we are interested in evacuation algorithms.

Sharing the bike. An interesting feature of our problem is the distinction between the hikers and the bike. On the one hand, the hikers are autonomous mobile agents that can move around on their own with speed 1 and communicate with each other. On the other hand, the bike is not autonomous and cannot move and/or communicate on its own and thus plays only the role of assistant in the search. The hiker using the bike has an advantage in that it can move with speed $v > 1$ which is of course faster than its walking speed 1. However, in our model the bike is also a limited resource in that it can be used by only one hiker at a time. This creates an interesting trade off for the evacuation time. The hikers would want to ride the bike to find the exit earlier. However, if the bike is not shared the evacuation time may get worse as the hiker not using the bike may worsen the overall evacuation time. This also implies that the hiker riding the bike has an advantage in sharing the bike with the other hiker as this will ultimately improve the overall evacuation time.

Bike Switching. An important aspect in our algorithms will be "bike switching", by which we mean changing the rider of the bike. We will assume throughout the paper that bike switching between hikers is instantaneous and at no time cost. Note that the hikers may recognize the presence of the bike when they are at the same location as the bike. From now on, to facilitate our discussions we will refer to the hiker riding the bike as the biker, which may be either of the two hikers.

Communication. A designated point on the infinite line is the exit and can be recognized as such by any of the hikers when they are at the same location as the exit.

The hikers may communicate throughout the execution of the algorithms. Two types of communication will be studied, namely wireless (also known as wifi) and face-to-face. In the former, the hikers can communicate instantaneously and at any distance, while in the latter only when they are at the same location and at the same time. The fact that a hiker is riding the bike does not diminish its ability to communicate. A typical communication exchange may involve, e.g., "exit is found", "bike released", "switch bike", etc. Note that the hikers are endowed with pedometers and have computing abilities so that they can deduce the location of the other hiker and/or the bike from relevant communications exchanged and/or the protocols they execute.

Notation. Throughout the paper we will be using R_1 and R_2 to denote the two hikers and B to denote the bike. The hikers are equipped with pedometers and are identical in all their capabilities (locomotion and communication) and the subscripts $i = 1, 2$ in R_i do not imply that the hikers have identifiers. The origin of the real line will be at the point $x = 0$ on the x-axis and this will also be the starting location of the hikers and the bike. The adversary may place the exit at either of the points $\pm d$, where $d > 0$ will denote the unknown distance of the exit from the origin. In addition, $v > 1$ will denote the speed that a biker can attain when riding the bike.

1.2. Related Work

The continuous infinite line is a widely-used search domain. It is in this particular domain that the first search problems in the literature were proposed in the seminal papers [2,3] with a focus on stochastic search models and their analysis. Influential research for deterministic search by a single robot in the infinite line was developed in the work of [1] by proving that searching for a target has competitive ratio equal to 9, and for randomized search on the star graph by [4].

Evacuation is a form of group search in which the robots need to cooperate so that they all find the exit. It arises as a natural problem on the infinite line for the case of robots with faults (crash and/or Byzantine). The two important papers are [5] for robots with crash faults and [6] for robots with Byzantine faults. The study of evacuation in distributed computing for a unit disk is also related and was initiated in the paper [7] for both the wireless and face-to-face models. The reader can find additional related work on the continuous search domain in the survey paper [8].

The addition of an immobile token to aid in the exploration has been considered in the context of the rendezvous problem on a ring [9]. An extension of this work to mobile tokens can be found in [10]. In both of these papers the token is passive and is merely being used as a marker for the presence of the most recent "visit" of another agent. Similarly, in [11] the authors consider searching for a non-adversarial, uncooperative agent, called bus, which is moving with constant speed along the perimeter of a cycle. A different related model was investigated in [12] in which during search a robot can encounter a point or a sequence of points enabling faster and faster movement and the main goal is to adopt the route which allows a robot to reach the destination as quickly as possible.

Two related papers are [13] and its followup journal version [14]. In the former paper the authors introduce evacuation on an infinite line in the F2F model for two robots having max speed 1 and prove that 9 is a tight bound for evacuation. In the followup paper [14] tight bounds are shown for two robots with different speeds in the F2F model. In their model the robots can vary their speed between the min and max value. However, unlike our model, the slower robot is never able to move at the speed of the faster robot. As a consequence in our model a "shared" bike has the effect of averaging the speeds and improving the overall evacuation time of the ensemble. The main idea considered in the present paper of bike assisted evacuation modelled as a passive agent that can enhance the robots' evacuation time has not been considered in the relevant literature on search and evacuation before. (The present study is revised and updated from the first author's MCS Thesis [15]. A preliminary study without proofs in [16]).

It is also worth mentioning a different line of research that has evolved in recent years, concerning bike sharing systems in complementing traditional public transportation to reduce traffic congestion and mitigate atmospheric pollution. As a consequence bike-sharing has grown explosively everywhere [17]. This has led to extensive technical literature on different aspects of the performance of bike based transportation systems. For example, Ref. [18] addresses uncertainty in resource availability, Ref. [19] considers bike utilization conflict, Ref. [20] studies system balance maintainance, Ref. [21] investigates the efficient operation of shared mobility systems, Ref. [22] studies balancing, and [23] proposes a spatio-temporal bicycle mobility model. Finally we mention the recent paper [24] which gives a polynomial time algorithm for the Bike Sharing problem that produces an arrival-time optimal schedule for bikers to travel across the interval.

1.3. Outline and Results of the Paper

Our main results in the Wireless model are presented in Section 3. We give three algorithms: in the first the robots move in opposite direction with max speed, in the second with a specially selected "optimal" speed, and in the third the hiker imitates the biker. Results on the Face-to-Face model are presented in Section 4. We give three algorithms: in the first algorithm the hiker pursues the biker, in the second the hiker and the biker use zig-zag algorithms with specially chosen expansion factors, and in the third the algorithm establishes a sequence of specially constructed meeting points near the exit. In either communication model we conclude that the optimal algorithm depends on the speed v of the bike which we also determine. Details of the results are in Table 1.

Table 1. Main algorithms presented in the paper in the WiFi (top three) and Face-to-Face (F2F) (bottom three) models, the theorem where the analysis, and their corresponding evacuation time as a function of the bike's speed v, where $v > 1$.

Algorithm	Theorem	Evacuation Time
Algorithm 2 (WiFi)	Theorem 1	$\max\left\{2d + \frac{d}{v}, \frac{2d}{v} + \frac{d}{2} + \frac{d}{2v^2}\right\}$
Algorithm 3 (WiFi)	Theorem 2	$\frac{3d + 3vd + d\sqrt{v^2 + 26v + 9}}{4v}$
Algorithm 4 (WiFi)	Theorem 3	$\frac{9d}{v} + \frac{d}{2} - \frac{d}{2v^2}$
Algorithm 5 (F2F)	Theorem 4	$\frac{9d}{v} + d - \frac{5d}{8v^2}$
Algorithm 6 (F2F)	Theorem 5	$\frac{3dv^3 + 63dv^2 + 15dv - 9d}{2v^2(3v + 1)}$
Algorithm 7 (F2F)	Theorem 6	$3d - \frac{5v^2 - 12v - 1}{2v(v - 1)}d + \frac{5v^2 - 12v - 1}{v(v - 1)(3v - 1)}d \quad \text{if } 1 < v \le \frac{6 + \sqrt{41}}{5}$ $3d - \frac{5v^2 - 12v - 1}{2v(v - 1)}d + \frac{4(5v^2 - 12v - 1)}{v(v - 1)(3v - 1)}d \quad \text{if } \frac{6 + \sqrt{41}}{5} \le v$

We also establish lower bounds in Section 5. The competitive ratio of the algorithms can be given by dividing by $\frac{v+1}{2v}d$, where d is the distance of the exit from the origin (see Theorem 7). The main motivation of our current study on robot evacuation from an unknown target is to better understand the effect that communication models (F2F and Wireless) have on search and evacuation time for an autonomous mobile agent which is aided by another mobile agent (bike) which has limited mobility capabilities.

2. Preliminaries

In this section we remind the reader that the competitive ratio for the Zig-Zag search algorithm is known to be at most 9. A canonical Zig-Zag search algorithm is defined as follows (Algorithm 1):

Algorithm 1: Zig-Zag Algorithm

Consider X as infinite sequence of distances $2^0, 2^1, ...$;
for $i \leftarrow 0$ **to** ∞ **do**
    if *i is odd(resp.even)* **then**
        Move right (resp. left) a distance 2^k unless the exit is found;
        if *exit is found* **then**
          | Quit search
        end
        Turn; then move left (resp. right), return to origin
    end
end

The competitive ratio for Algorithm 1 is calculated as follows:

Every time the robot changes direction and moves twice the previous distance. Thus, if the exit is at distance d, then $2^k < d \leq 2^{k+2}$ for some k. Hence the search time T will be calculated as follows:

$$T = 2 \cdot 1 + 2 \cdot 2 + \cdots + 2 \cdot 2^{k+1} + d$$
$$= 2 \cdot (2^{k+2} - 1) + d = 2^3 \cdot 2^k - 2 + d \leq 8d + d = 9d$$

Thus, the competitive ratio of this algorithm is $\frac{9d}{d} = 9$. The lower bound proof can be found in [1].

3. Evacuation in the Wireless (WiFi) Model

In this section we provide our main algorithms in the wireless communication model. In this model the two hikers can communicate instantaneously at any distance. Three algorithms will be considered and analyzed their evacuation time depends on the maximum speed v of the biker.

3.1. Opposite Direction with Max Speed

In Algorithm 2, the hiker and the biker move in opposite directions with their maximum speed, assuming that the biker moves with speed v. The one which finds the exit first will communicate with the other to proceed to the exit. Moreover, if it is the biker that found the exit first it returns the bike to a suitable position and shares it with the hiker. Details of the algorithm are as follows.

Algorithm 2: (OppDirectionWithMaxSpeedWiFi)

The hiker and the biker move in opposite directions. The one that finds the exit
    first communicates it to the other;
if *the hiker found the exit first* **then**
    | the biker moves to the exit at full speed;
end
else if *the biker found the exit first* **then**
    | it returns and drops the bike off to an appropriately chosen position x and
      shares the bike with the hiker;
end
Stop when they both arrive at the exit;

Theorem 1. *The evacuation time for Algorithm 2 in the WiFi model is at most* $2d + \frac{d}{v}$.

Proof of Theorem 1. Without loss of generality assume the hiker moves in the leftward and the biker in the rightward direction both starting from the origin. There are two cases to consider depending on whether the hiker or the biker finds the exit first.

Case 1: The exit is found by the biker.

When the biker finds the exit, which is at distance d from the origin, it has spent time $\frac{d}{v}$. The biker will communicate with the hiker, to come to the exit. Since the exit is found by the biker first, the biker can help the hiker by returning and dropping off the bike at some distance x away from the exit such that the hiker can pick it up and arrive to the exit at the same time as the other robot. It is easy to see that to find the distance x where the bike is dropped off, we need to solve the equation $\frac{x}{v} + x = d + \frac{d}{v} - x + \frac{x}{v}$. This leads to the solution $x = \frac{d}{2} + \frac{d}{2v}$. Hence the hiker which is at distance $\frac{d}{v}$ when the biker reaches the exit, will need $\frac{d}{v}$ to reach the origin, in addition to $d - x + \frac{x}{v} = \frac{d}{2} + \frac{d}{2v^2}$ to reach the exit. Therefore, the evacuation time will be $\frac{2d}{v} + \frac{d}{2} + \frac{d}{2v^2}$.

Case 2: The exit is found by the hiker.

When the hiker finds the exit which is at distance d to the left of the origin, the biker will be at distance dv to the right of the origin. The hiker will communicate with the biker to come to the exit. The biker turns back and goes to the exit which takes additional time $d + \frac{d}{v}$. It follows that the evacuation time in this case will be $d + d + \frac{d}{v} = 2d + \frac{d}{v}$.

Therefore, by combining the two cases above, we conclude that the evacuation time for this algorithm will be $\max\left\{2d + \frac{d}{v}, \frac{2d}{v} + \frac{d}{2} + \frac{d}{2v^2}\right\}$. This completes the proof of Theorem 1. $\square$

3.2. Opposite Direction with Optimal Speed

Unlike Algorithm 2 in which the hiker and biker use their maximum speed, in the next Algorithm 3 the biker will not use its maximum speed v. Instead it will move with a specially chosen speed u which is less than v. The hiker or the biker which finds the exit first will communicate with the other which will then move towards the exit with its maximum speed. It turns out that the modified algorithm performs better than the previous one and it is optimal up to a certain maximum speed v. Assuming that R_1 is the hiker and R_2 is the biker, the algorithm will be as follows.

Algorithm 3: (OppDirectionWithOptimalSpeedWiFi)

R_1 moves left with unit speed;
R_2 moves right with speed $u = \frac{1}{4}(-v - 1 + \sqrt{v^2 + 26v + 9})$;
if R_1 *reaches the exit* **then**
    Inform R_2 about the location of the exit;
    R_2 moves toward the exit with its maximum speed v;
end
else if R_2 *reaches the exit* **then**
    Inform R_1 about the location of the exit;
    Drop-off the bike at distance $\frac{d}{2} + \frac{d}{2u}$;
    R_2 heads back toward the exit;
    R_1 reverses the direction back to the exit then picks up the bike and moves to the exit with maximum speed v;
end

Theorem 2. *The evacuation time for Algorithm 3 using the WiFi model is at most* $\frac{3d + 3vd + d\sqrt{v^2 + 26v + 9}}{4v}$.

Proof of Theorem 2. Assume that R_1 which is the hiker moves in the leftward direction and R_2, which is the biker, moves in the rightward direction. There are two cases to consider depending on who reaches the exit first.

Case 1: Hiker reaches the exit first.

The time needed by R_1 to reach the exit is d. At this point R_2 will be at distance du away from the origin since it is moving with speed u. As mentioned in the algorithm, R_2 will use its maximum speed v on the way back. Thus, it needs $\frac{du}{v}$ to reach the origin and it will need additional time $\frac{d}{v}$ to join R_1 and reach the exit. Therefore, the evacuation time in this case will be

$$T_1 = d + \frac{du}{v} + \frac{d}{v} = \frac{dv + du + d}{v}$$

Case 2: Biker reaches the exit first.

The time needed by R_2 to reach the exit is $\frac{d}{u}$. As soon as R_2 reaches the exit, it will inform R_1 immediately. At this point in time, R_1 will be at distance $\frac{d}{u}$ on the other side of the origin since it is moving with unit speed. R_2 will go back distance x to drop off the bike for R_1. The key point to find x is to have R_2 drop off the bike in a way that R_1 can pick it up and arrive at the same time as R_2. We know from Algorithm 2 that R_2 will not use its maximum speed and will move with speed u only until it reaches the exit. The only reason for not using its maximum speed before reaching the exit is to avoid having R_1 and R_2 farther apart from each other since this will increase the overall evacuation time. Thus when R_2 goes back to drop off the bike, it will use its maximum speed v. Based on that we have the following equation.

$$x + \frac{x}{v} = d + \frac{d}{u} - x + \frac{x}{v}$$

whose solution is $x = \frac{d}{2} + \frac{d}{2u}$. Substituting x in order to calculate the evacuation time T_2 yields:

$$\begin{aligned}
T_2 &= \frac{d}{u} + x + \frac{x}{v} \\
&= \frac{d}{u} + \frac{d}{2} + \frac{d}{2u} + \frac{d}{2v} + \frac{d}{2uv} \\
&= \frac{2dv + duv + dv + du + d}{2uv} \\
&= \frac{3dv + du + duv + d}{2uv}
\end{aligned}$$

In order to find the best evacuation time, we need to find the best value of u which makes the maximum of T_1 and T_2 minimized given that $1 \leq u \leq v$. This means that the objective is to minimize the following quantity

$$\max\{T_2, T_2\} = \max\left\{ \frac{dv + du + d}{v}, \frac{3dv + du + duv + d}{2uv} \right\} \tag{1}$$

In order to find the solution for (1), we determine the point of intersection of the evacuation time plots for T_1 and T_2. This will give the following:

$$\frac{u + v + 1}{v} = \frac{3v + u + uv + 1}{2uv} \implies 2u^2v + 2uv^2 + 2uv = 3v^2 + uv + uv^2 + v$$

$$\implies 2u^2v + (v^2 + v)u - 3v^2 - v = 0$$

The last equation will have two roots, and choosing the positive one gives the following solution for u:

$$u = \frac{1}{4v}\left(-v^2 - v + \sqrt{v^4 + 2v^3 + v^2 - 8v(-3v^2 - v)}\right)$$

$$= \frac{1}{4v}\left(-v^2 - v + \sqrt{v^4 + 2v^3 + v^2 + 24v^3 + 8v^2}\right)$$

$$= \frac{1}{4v}\left(-v^2 - v + \sqrt{v^4 + 26v^3 + 9v^2}\right)$$

$$= \frac{1}{4}\left(-v - 1 + \sqrt{v^2 + 26v + 9}\right)$$

In order to get the evacuation time T, we can substitute u in T_1 or T_2, we get the following:

$$T = \frac{dv + du + d}{v}$$

$$= \frac{1}{v}\left(dv + \frac{-dv - d + d\sqrt{v^2 + 26v + 9}}{4} + d\right)$$

$$= \frac{4dv - dv - d + d\sqrt{v^2 + 26v + 9} + 4d}{4v}$$

$$= \frac{3d + 3dv + d\sqrt{v^2 + 26v + 9}}{4v}.$$

This completes the proof of Theorem 2. $\square$

3.3. Slower Imitates Faster

In the next Algorithm 4 the robots perform a "doubling zig-zag" strategy with different parameters. The biker is using a doubling strategy to search for the exit and moves a distance 2^k during the k-th iteration. The hiker is also using a doubling strategy but since it is moving with unit speed it will try to stay as close as possible to the biker. This can be achieved by having the hiker move a distance $\frac{2^k}{v}$ during the k-th iteration, since moving any further will cause the hiker to be farther away from the biker during the $(k + 1)$-st iteration. Assuming that robot R_1 is the biker and robot R_2 is the hiker, the algorithm will be as follows.

Algorithm 4: (SlowerImitateFasterWiFi)

---

for $k \leftarrow 1$ **to** ∞ **do**

    if *k is odd (resp.even)* **then**

        R_1 moves right (resp. left) a distance 2^k unless the exit is found;

        R_2 moves right (resp. left) a distance $\frac{2^k}{v}$;

        if *exit is found by* R_1 **then**

            Communicate with R_2;

            R_1 moves back $\frac{d}{2} - \frac{d}{2v}$ to leave the bike for R_2 and then returns to exit;

            R_2 continues toward the exit after picking up the bike left by R_1;

            Quit;

        end

        R_1 turns; then moves left (resp. right), returns to the origin;

        R_2 turns; then moves left (resp. right), returns to the origin;

    end

end

Theorem 3. *The evacuation time for Algorithm 4 using the WiFi model is at most* $\frac{9d}{v} + \frac{d}{2} - \frac{d}{2v^2}$.

Proof of Theorem 3. In this algorithm the biker uses a doubling strategy with maximum speed v. The hiker will follow the biker but will move $\frac{2^k}{v}$ in each iteration instead of 2^k. The biker will reach the exit first then will communicate with the hiker to proceed to the exit. The biker will go back distance $\frac{d}{2} - \frac{d}{2v}$ to drop off the bike so that the hiker can pick it up on its way to the exit. We will justify why biker R_1 needs to move $\frac{d}{2} - \frac{d}{2v}$ after reaching the exit to leave the bike for hiker R_2.

After biker R_1 reaches the exit, there is no benefit to stay at the exit with the bike since hiker R_2 which is moving with unit speed can benefit from the bike to reach the exit faster. The key to find the distance x which is the distance between the exit and the point where the bike is dropped off is to have biker R_1, drop it off at a point such that when it goes back to the exit it will reach the exit at the same time as hiker R_2. If we consider that d is the distance from the origin to the exit and x is the distance from the exit to the point where biker R_1 drops off the bike, then we have $d - x + \frac{x}{v} = \frac{d}{v} + \frac{x}{v} + x$ which leads to $x = \frac{d}{2} - \frac{d}{2v}$. This will guarantee that when the biker drops off the bike at distance x, it will reach the exit at the same time as the hiker. Hence we guarantee that the bike is not kept unnecessarily with the robot which reaches the exit first.

Assume that the exit is found during the kth iteration, then $2^{k-2} < d \leq 2^k$. We can calculate the evacuation time as follows:

$$
\begin{aligned}
T &= \frac{2 \cdot 2^0}{v} + \frac{2 \cdot 2^1}{v} + \cdots + \frac{2 \cdot 2^{k-1}}{v} + d - x + \frac{x}{v} \\
&= \frac{2(2^k - 1)}{v} + d - x + \frac{x}{v} \\
&= \frac{2^{k+1}}{v} - \frac{2}{v} + d - \frac{d}{2} + \frac{d}{2v} + \frac{d}{2v} - \frac{d}{2v^2} \\
&= \frac{2^{k+1}}{v} - \frac{2}{v} + \frac{d}{2} + \frac{d}{v} - \frac{d}{2v^2} \\
&\leq 2^3 \cdot \frac{2^{k-2}}{v} - \frac{2}{v} + \frac{d}{2} + \frac{d}{v} - \frac{d}{2v^2} \\
&\leq \frac{8d}{v} - \frac{2}{v} + \frac{d}{2} + \frac{d}{v} - \frac{d}{2v^2} \\
&\leq \frac{9d}{v} + \frac{d}{2} - \frac{d}{2v^2} - \frac{2}{v} \\
&\leq \frac{9d}{v} + \frac{d}{2} - \frac{d}{2v^2}
\end{aligned}
$$

This completes the proof of Theorem 3. $\square$

Figure 1 depicts and compares the performance of the three algorithms presented for the WiFi model.

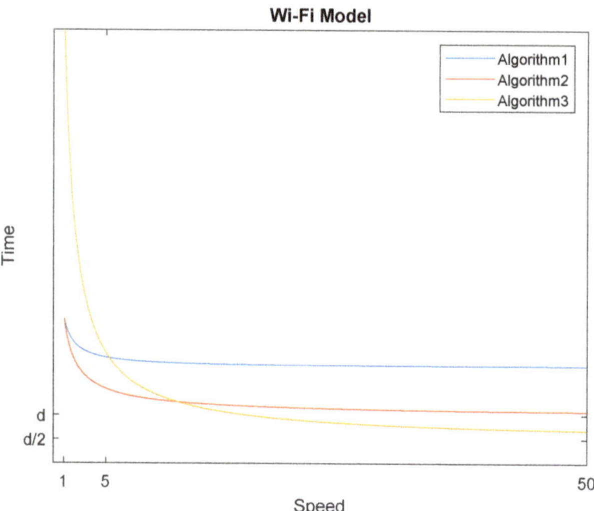

Figure 1. Graph for the three algorithms using the wi-fi model. On high speed, the evacuation time for Algorithm 4 converges to $\frac{d}{2}$ versus $2d$ and d for Algorithms 2 and 3, respectively.

4. Evacuation in the Face-to-Face (F2F) Model

In this section we provide our main algorithms in the face-to-face communication model. Recall that in this model the hikers can exchange messages only if they occupy the same location at the same time.

4.1. Slower Pursues Faster

In the first Algorithm 5, we assume that the hiker will follow the biker. Since the biker is using a "doubling zig-zag" strategy, during any iteration, let us say the k-th one, the biker will reverse the direction after reaching 2^k and will meet the hiker at some point X_k. At the meeting point the hiker will reverse its direction. We notice from this that the hiker is following a deterministic strategy specified through a sequence of points $X_1, X_2, \ldots, X_k, \ldots$ (to be defined later), where each X_k represents the meeting point for the hiker and the biker during the k-th iteration. In other words, the biker will follow a doubling strategy with factor 2^k while the hiker will follow the sequence $X_1, X_2, \ldots, X_k, \ldots$ above. When the biker reaches the exit, it will go back a certain distance x, which will be determined later, to drop off the bike and then will return back walking toward the exit.

In the algorithm below we use the parameters $a := \frac{1-v}{1+v}$, $b := \frac{1}{1+v}$. Further, we assume that R_1 is the biker and R_2 is the hiker.

Theorem 4. *The evacuation time for Algorithm 5 using the F2F model is at most $\frac{9d}{v} + d - \frac{5d}{8v^2}$.*

Proof of Theorem 4. In order to find the sequence $\{X_1, X_2 \ldots X_k\}$, we argue as follows. During the 1st iteration, in order to calculate X_1, we know that the biker will move 2^0 to reach the peak point and then will come back $2^0 - X_1$ with speed v to reach point X_1 while the hiker will move X_1 with unit speed during the same time. Given that $a = \frac{1-v}{1+v}$ and $b = \frac{1}{1+v}$, we have the following:

$$X_1 = \frac{2^0}{v} + \frac{2^0}{v} - \frac{X_1}{v} \implies X_1 = \frac{2}{1+v} = 2 \cdot b$$

Algorithm 5: (SlowerPursueFasterF2F)

---

for $k \leftarrow 1$ **to** ∞ **do**
    if *k is odd(resp.even)* **then**
        R_1 moves right (resp. left) a distance 2^k unless the exit is found;
        R_2 moves right (resp. left) a distance $\frac{2b(2^k - a^k)}{2-a}$;
        if *the exit is found by R_1* **then**
            Move back distance x to drop off the bike for R_2 then switch direction toward the exit ;
            R_2 picks up bike and moves toward the exit;
            Quit;
        end
        R_1 turns; then moves left (resp. right), return to origin;
        R_2 turns; then moves left (resp. right), return to origin;
    end
end

During the 2nd iteration we have the following:

$$X_1 + X_2 = \frac{1}{v}(X_1 + 2 + 2 - X_2)$$

$$\implies X_2 = \frac{4}{1+v} + \frac{1-v}{1+v} \cdot X_1 = a \cdot X_1 + 2^2 \cdot b$$

Since $X_2 = a \cdot X_1 + 2^2 \cdot b$, then for the kth iteration we have:

$$X_k = a \cdot X_{k-1} + 2^k \cdot b$$

Replacing $X_{k-1} = a \cdot X_{k-2} + b \cdot 2^{k-1}$ in the above equation gives

$$X_k = a(a \cdot X_{k-2} + b \cdot 2^{k-1}) + 2^k \cdot b$$

Similarly replacing $X_{k-2} = a \cdot X_{k-3} + b \cdot 2^{k-2}$ gives

$$X_k = a^3 \cdot X_{k-3} + a^2 \cdot 2^{k-2} \cdot b + a \cdot b \cdot 2^{k-1} + 2^k \cdot b$$

Recursing down to X_1 leads to the following calculation:

$$X_k = b \cdot 2^k \left(\left(\frac{a}{2}\right)^0 + \cdots + \left(\frac{a}{2}\right)^{k-1} \right)$$

$$= \frac{b \cdot 2^k (1 - (\frac{a}{2})^k)}{1 - \frac{a}{2}}$$

$$= \frac{2 \cdot b (2^k - a^k)}{2^k(2-a)} \cdot 2^k \cdot b$$

$$= \frac{2 \cdot b (2^k - a^k)}{2-a}$$

Assume that the exit is found during the kth iteration of the algorithm. Before writing down the evacuation time, let us find the distance x away from the exit, where the bike will be dropped off by the biker. We know that the biker and the hiker will meet at each entry of the sequence and eventually they will meet at X_{k-1}.

- Define T_1 to be the time needed by the biker to go to the exit from the point of intersection between the the biker and the hiker at X_{k-1} then to return distance x to drop off the bike and subsequently go back to the exit. Thus T_1 can be defined as follows:

$$T_1 = \frac{1}{v}(X_{k-1} + x + d) + x.$$

- Define T_2 to be the time needed by the hiker to go from the point of intersection between the hiker and the biker at X_{k-1} to the exit while picking up the bike on its way. Thus T_2 can be defined as follows:

$$T_2 = X_{k-1} + d - x + \frac{x}{v}.$$

The best thing that the biker can do is to drop off the bike and arrive at the same time to the exit with the hiker who will pick up the bike on its way. This can be achieved by having $T_1 = T_2$. In turn, this yields

$$\frac{1}{v}(X_{k-1} + x + d) + x = X_{k-1} + d - x + \frac{x}{v}$$

$$\implies x = \frac{1}{2}(X_{k-1} + d) - \frac{1}{2v}(d + X_{k-1})$$

Thus, the evacuation time T can be written as follows:

$$T = \frac{1}{v}(2 \cdot 2^0 + 2 \cdot 2^1 + \cdots + 2 \cdot 2^{k-1}) + \frac{d}{v} + \frac{x}{v} + x$$

Replacing the value of x which was calculated above gives:

$$T = \frac{2}{v}(2^k - 1) +$$
$$\frac{d}{v} + \frac{d}{2v} - \frac{d}{2v^2} + \frac{d}{2} - \frac{d}{2v} + \frac{1}{2v}X_{k-1} - \frac{1}{2v}X_{k-1} + \frac{1}{2}X_{k-1} - \frac{1}{2v^2}X_{k-1}$$
$$= \frac{2^{k+1}}{v} - \frac{2}{v} + \frac{d}{v} + \frac{d}{2} - \frac{d}{2v^2} + \left(\frac{1}{2} - \frac{1}{2v^2}\right)X_{k-1}$$

Now recall that $X_{k-1} = \frac{2b(2^{k-1} - a^{k-1})}{2 - a}$, where $a = \frac{1-v}{1+v}$ and $b = \frac{1}{1+v}$. Since $-1 < a < 0$ and $0 < b \leq \frac{1}{2}$ it is obvious that $X_{k-1} \leq \frac{1}{2}(2^{k-1} + 1) = 2^{k-2} + \frac{1}{2}$. Moreover, knowing that $2^{k-2} < d \leq 2^k$, the evacuation time T can be simplified as follows:

$$T \leq \frac{2^{k+1}}{v} - \frac{2}{v} + \frac{d}{v} + \frac{d}{2} - \frac{d}{2v^2} + \left(\frac{1}{2} - \frac{1}{2v^2}\right)\left(2^{k-2} + \frac{1}{2}\right)$$
$$\leq \frac{2^{k+1}}{v} - \frac{2}{v} + \frac{d}{v} + \frac{d}{2} - \frac{d}{2v^2} + 2^{k-3} + \frac{1}{4} - \frac{2^{k-2}}{2v^2} - \frac{1}{4v^2}$$
$$\leq \frac{8d}{v} - \frac{2}{v} + \frac{d}{v} + \frac{d}{2} - \frac{d}{2v^2} + \frac{d}{2} + \frac{1}{4} - \frac{d}{8v^2} - \frac{1}{4v^2}$$
$$\leq \frac{9d}{v} + d - \frac{5d}{8v^2} - \frac{1}{4v^2} + \frac{1}{4} - \frac{2}{v}$$
$$\leq \frac{9d}{v} + d - \frac{5d}{8v^2}$$

This completes the proof of Theorem 4. $\square$

4.2. Slower Evacuation Close to Exit without Aid

In the second Algorithm 6, the biker uses a "doubling zig-zag" strategy with its maximum speed v, while the hiker will try to be as close as possible to the biker. In order to achieve that, the hiker will use a doubling strategy as well but will use its own "expansion" factor. The factor will be determined based on the fact that both the hiker and the biker should meet at a specific point during each iteration. These meeting points will form a sequence whose k-th element during iteration k is taken to be equal to $\frac{2^{k+2}}{3v+1}$. During the last iteration, when the biker finds the exit, the hiker will eventually reach the meeting point and will not find the biker there. This will let it know that it should keep going toward the exit. Assuming that initially R_1 is the biker and R_2 is the hiker, the algorithm will be as follows:

Algorithm 6: (SlowerEvacuationCloseToExitWithoutAidF2F)

---

for $k \leftarrow 1$ **to** ∞ **do**

    if *k is odd (resp.even)* **then**

        R_1 moves right (resp. left) a distance 2^k unless the exit is found;

        R_2 moves right (resp. left) a distance $\frac{2^{k+2}}{3v+1}$;

    if *k=1* **then**

        R_2 waits for R_1;

    end

    if *exit is found by* R_1 **then**

        Move back distance $\frac{d}{2} - \frac{d}{2v} + \frac{2^k(v-1)}{v(3v+1)}$ to drop off the bike for R_2 then

        switch direction toward the exit;

        R_2 picks up bike and moves toward exit;

        Quit;

    end

    R_1 turns; then moves left (resp. right), return to origin;

    R_2 turns; then moves left (resp. right), return to origin;

end

end

Theorem 5. *The evacuation time for Algorithm 6 using the F2F model is at most*
$$\frac{3dv^3 + 63dv^2 + 15dv - 9d}{2v^2(3v+1)}.$$

Proof of Theorem 5. The biker is using a doubling strategy and is moving 2^k during each iteration k. The hiker will use a doubling strategy as well and it will follow the biker. In order to keep the hiker as close as possible to the biker, we must find the sequence that the hiker should follow. We assume that both the hiker and the biker meet at a certain point X_k and that they are willing to meet at point X_{k+1} at the same time without waiting for one another, taking into consideration that $X_{k+1} = 2X_k$. The sequence can be calculated as follows:

$$X_k + X_{k+1} = \frac{1}{v}(X_k + 2^{k+1} + 2^{k+1} - X_{k+1})$$

$$\implies X_k + 2X_k = \frac{1}{v}(X_k + 2^{k+1} + 2^{k+1} - 2X_k)$$

$$\implies \frac{3v+1}{v}X_k = \frac{2^{k+2}}{v}$$

$$\implies X_k = \frac{2^{k+2}}{3v+1}$$

We have the sequence $\{X_0, X_1, \ldots, X_k\}$ given that $X_k = \frac{2^{k+2}}{3v+1}$ where $k \geq 1$. Each of the hiker and the biker will use its own doubling strategy. During each iteration, they will meet on both sides at specific points which are elements of the above sequence. During the kth iteration, when the biker reaches the exit, it will move back distance x to drop off the bike such that the hiker can pick it up and reach the exit at the same time as itself. The distance x can be calculated as follows: $\frac{d}{v} + \frac{1}{v} \cdot X_{k-1} + \frac{x}{v} + x = d + X_{k-1} - x + \frac{x}{v}$. Substituting X_{k-1} this becomes $\frac{d}{v} + \frac{2^{k+1}}{v(3v+1)} + \frac{x}{v} + x = d + \frac{2^{k+1}}{3v+1} - x + \frac{x}{v}$ Solving for x, the last equation yields $x = \frac{d}{2} - \frac{d}{2v} + \frac{2^k(v-1)}{v(3v+1)}$. Assuming that $2^{k-2} < d \leq 2^k$ and replacing x which was calculated above, the evacuation time T can be computed as follows:

$$\begin{aligned}
T &= \frac{1}{v}(2 \cdot 2^0 + 2 \cdot 2^1 + \cdots + 2 \cdot 2^{k-1}) + \frac{d}{v} + \frac{x}{v} + x \\
&= \frac{2}{v}(2^k - 1) + \frac{d}{v} + \frac{x}{v} + x \\
&= \frac{2^{k+1}}{v} - \frac{2}{v} + \frac{d}{v} + \frac{x}{v} + x \\
&= \frac{2^{k+1}}{v} - \frac{2}{v} + \frac{d}{v} + \frac{d}{2v} - \frac{d}{2v^2} + \frac{2^k(v-1)}{v^2(3v+1)} + \frac{d}{2} - \frac{d}{2v} + \frac{2^k(v-1)}{v(3v+1)} \\
&\leq \frac{16d}{2v} - \frac{2}{v} + \frac{d}{v} + \frac{d}{2} - \frac{d}{2v^2} + \frac{4d(v-1)}{v^2(3v+1)} + \frac{4d(v-1)}{v(3v+1)} \\
&\leq \frac{18d}{2v} + \frac{d}{2} - \frac{d}{2v^2} + \frac{4d(v-1)}{v^2(3v+1)} + \frac{4d(v-1)}{v(3v+1)} - \frac{2}{v} \\
&\leq \frac{54dv^2 + 18dv + 3dv^3 + dv^2 - 3dv - d + 8vd - 8d + 8dv^2 - 8dv}{2v^2(3v+1)} - \frac{2}{v} \\
&\leq \frac{3dv^3 + 63dv^2 + 15dv - 9d}{2v^2(3v+1)} - \frac{2}{v} \\
&\leq \frac{3dv^3 + 63dv^2 + 15dv - 9d}{2v^2(3v+1)}
\end{aligned}$$

This completes the proof of Theorem 5. $\square$

4.3. Nearest Meeting to Exit

In order to reduce the evacuation time, it is more suitable for the biker to search for the exit while the hiker follows a "doubling zig-zag" strategy that will keep it as close as possible to the biker and will expedite its travel time to the exit during the last iteration of the evacuation algorithm. In order to achieve that, the purpose of the next algorithm will be to find this deterministic doubling strategy that the hiker should follow. Assuming that R_1 is the biker and R_2 is the hiker, Algorithm 7 will be as follows.

Algorithm 7: (EvacuatingWithBikeF2F)

---

for $k \leftarrow 1$ **to** ∞ **do**

    if *k is odd (resp.even)* **then**

        R_1 moves right (resp. left) a distance 2^k unless the exit is found;

        R_2 moves right (resp. left) a distance $\frac{2^{k+1}}{3v-1}$;

        if $k = 1$ **then**

            R_2 waits for R_1;

        if *exit is found by R_1* **then**

            R_1 switches direction to inform R_2;

            R_1 drops off the bike at distance $\frac{x}{2}$;

            R_2 picks up the bike and continues to the exit;

            Quit;

        if *exit is found by R_2* **then**

            R_2 waits till R_1 comes to the exit;

            Break;

        R_1 turns; then moves left (resp. right), return to origin;

        R_2 turns; then moves left (resp. right), return to origin;

Theorem 6. *The evacuation time for Algorithm 7 using the F2F model is upper bounded by*

$$3d - \frac{5v^2 - 12v - 1}{2v(v-1)}d + \frac{5v^2 - 12v - 1}{v(v-1)(3v-1)}d \quad if\, 1 < v \le \frac{6+\sqrt{41}}{5}$$

$$3d - \frac{5v^2 - 12v - 1}{2v(v-1)}d + \frac{4(5v^2 - 12v - 1)}{v(v-1)(3v-1)}d \quad if\, \frac{6+\sqrt{41}}{5} \le v$$

Proof of Theorem 6. Let us consider the sequence: $X = \{X_1, X_2, X_3, \ldots, X_k\}$, where $X_k = r \cdot X_{k-1}$. The purpose is to find the best value of r which is the factor related to the doubling strategy that the hiker follows. Definitely the best meeting point would be the peak point reached by the hiker during each iteration, since it will be the closest to the exit. Assume that both the hiker and the biker meet at some point X_{k-1} during the $k-1$ iteration and they are willing to meet during the k-th iteration without waiting for one another, then we have:

$$X_{k-1} + X_k = \frac{1}{v} \cdot (2^{k-1} - X_{k-1}) + \frac{2^{k-1}}{v} + \frac{X_k}{v}$$

and after substituting X_k

$$X_{k-1} + r \cdot X_{k-1} = \frac{1}{v} \cdot (2^{k-1} - X_{k-1}) + \frac{2^{k-1}}{v} + \frac{r}{v} \cdot X_{k-1}.$$

In turn, this implies

$$(r+1) \cdot X_{k-1} = \frac{2^k}{v} + \frac{r-1}{v} \cdot X_{k-1}$$

$$\implies r \cdot v \cdot X_{k-1} + v \cdot X_{k-1} - r \cdot X_{k-1} + X_{k-1} = 2^k$$

$$\implies X_{k-1} = \frac{2^k}{r \cdot v + v - r + 1}$$

Similarly we have $X_k = \frac{2^{k+1}}{r \cdot v + v - r + 1}$. Consider $X_k = r \cdot X_{k-1}$, then we can deduce that $r = 2$. Substituting $r = 2$ gives $X_k = \frac{2^{k+1}}{3v-1}$. So we conclude that the hiker will use doubling strategy and will follow the sequence $X = \{\frac{4}{3v-1}, \frac{8}{3v-1}, \ldots, \frac{2^{k+1}}{3v-1}\}$. Assume that R_1 is the biker which moves with speed v and R_2 is the hiker which moves with unit speed. Consider $d = \frac{2^{k+1}}{3v-1} + e$ where $e \ge 0$. Definitely R_1 will reach the exit before R_2. Since the exit is at distance e from the meeting point, then from that point on, R_1 needs time $\frac{e}{v}$ to reach the exit. During this time, R_2 will be at distance e on the other side of the meeting point. Therefore, when R_1 reaches the exit, R_2 will be at distance $e + \frac{e}{v} = e\frac{v+1}{v}$ away from the exit. The distance z that hiker R_2 moves from the point where R_1 reaches the exit till the point it meets R_1 will be as follows:

$$z = \frac{e(v+1)}{v^2} + \frac{z}{v}$$

$$\implies \frac{z(v-1)}{v} = \frac{e(v+1)}{v^2}$$

$$\implies z = \frac{v+1}{v-1} \cdot \frac{e}{v}$$

Therefore, when biker R_1 reaches hiker R_2 to inform it about the exit, R_2 will be far from the exit by a distance $x = \frac{v+1}{v-1} \cdot \frac{e}{v} + e\frac{v+1}{v}$. Now it is required to find at what distance y away from the exit should biker R_1 drop off the bike so that hiker R_2 can pick it up and proceed to the exit and reach it at the same time as R_1. After we find out the distance y, we will go back to create the algorithm for the two hikers with a bike model. In order to find out the distance y we have the following:

$$\frac{x}{v} - \frac{y}{v} + y = x - y + \frac{y}{v}$$
$$\implies \frac{x}{v} - \frac{y}{v} + 2y = x + \frac{y}{v}$$
$$\implies y(2v - 2) = x(v - 1)$$
$$\implies y = \frac{x(v-1)}{2(v-1)} = \frac{x}{2}$$

The evacuation time T can be calculated as follows:

$$T = \frac{2 \cdot 2^0 + 2 \cdot 2^1 + \cdots + 2 \cdot 2^{k-1}}{v} + \frac{d}{v} + \frac{x}{v} + \frac{y}{v} + y$$
$$= \frac{2(2^k - 1)}{v} + \frac{d}{v} + \frac{x}{v} + \frac{x}{2v} + \frac{x}{2}$$
$$= \frac{2^{k+1}}{v} - \frac{2}{v} + \frac{d}{v} + \frac{3}{2v}\left(e + \frac{e}{v} + \frac{v+1}{v-1} \cdot \frac{e}{v}\right) + \frac{1}{2}\left(e + \frac{e}{v} + \frac{v+1}{v-1} \cdot \frac{e}{v}\right)$$

Since $d = e + \frac{2^{k+1}}{3v-1}$, replacing $\frac{2^{k+1}}{v} = 3d - 3e - \frac{d}{v} + \frac{e}{v}$ in the above equation gives the following:

$$T = 3d - \frac{d}{v} - 3e + \frac{e}{v} - \frac{2}{v} + \frac{d}{v} + \frac{3e}{2v} + \frac{3e}{2v^2} + \frac{3e(v+1)}{2v^2(v-1)} + \frac{e}{2} + \frac{e}{2v} + \frac{e(v+1)}{2v(v-1)}$$
$$= 3d - \frac{2}{v} + \frac{3e}{v} - \frac{5e}{2} + \frac{3e}{2v^2} + \frac{3ev + 3e + ev^2 + ev}{2v^2(v-1)}$$
$$= 3d - \frac{2}{v} + \frac{3e}{v} - \frac{5e}{2} + \frac{3e}{2v^2} + \frac{ev^2 + 4ev + 3e}{2v^2(v-1)}$$
$$= 3d - \frac{2}{v} + e \cdot \frac{6v^2 - 6v - 5v^3 + 5v^2 + 3v - 3 + v^2 + 4v + 3}{2(v-1)v^2} =$$
$$= 3d - \frac{2}{v} - e \cdot \frac{5v^2 - 12v - 1}{2v(v-1)}$$
$$= 3d - \frac{2}{v} - \frac{5v^2 - 12v - 1}{2v(v-1)}\left(d - \frac{2^{k+1}}{3v-1}\right)$$
$$= 3d - d \cdot \frac{5v^2 - 12v - 1}{2v(v-1)} + 2^k \frac{5v^2 - 12v - 1}{v(v-1)(3v-1)} - \frac{2}{v}$$

There are two cases to consider here:

Case 1: $1 < v \le \frac{6+\sqrt{41}}{5}$.

Since $5v^2 - 12v - 1 \le 0$ and $d \le 2^k$ we have that

$$T \le 3d - \frac{5v^2 - 12v - 1}{2v(v-1)}d + \frac{5v^2 - 12v - 1}{v(v-1)(3v-1)}d - \frac{2}{v}$$
$$\le 3d - \frac{5v^2 - 12v - 1}{2v(v-1)}d + \frac{5v^2 - 12v - 1}{v(v-1)(3v-1)}d$$

Case 2: $\frac{6+\sqrt{41}}{5} \le v$.

Since $0 \le 5v^2 - 12v - 1$ and $2^{k-2} \le d$ we conclude that

$$T \le 3d - \frac{5v^2 - 12v - 1}{2v(v-1)}d + \frac{5v^2 - 12v - 1}{v(v-1)(3v-1)} \cdot 2^2 \cdot 2^{k-2} - \frac{2}{v}$$

$$\le 3d - \frac{5v^2 - 12v - 1}{2v(v-1)}d + \frac{4(5v^2 - 12v - 1)}{v(v-1)(3v-1)}d - \frac{2}{v}$$

$$\le 3d - \frac{5v^2 - 12v - 1}{2v(v-1)}d + \frac{4(5v^2 - 12v - 1)}{v(v-1)(3v-1)}d$$

This completes the proof of Theorem 6. $\square$

Figure 2 depicts and compares the performance of the three algorithms presented for the face-to-face model.

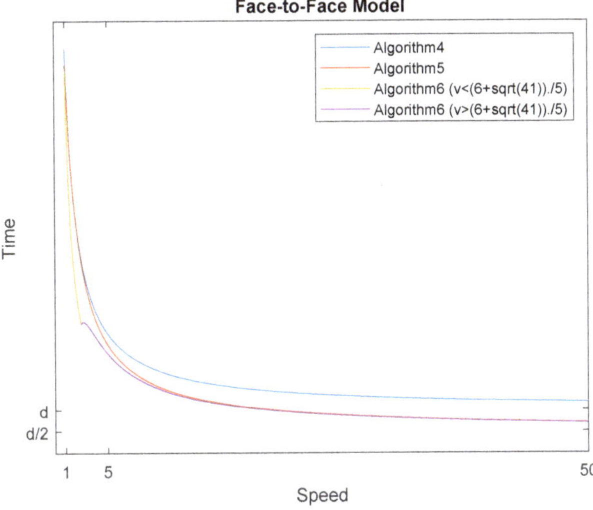

Figure 2. Graph for the three algorithms using the face-to-face model. On high speed, the evacuation time for Algorithm 7 converges to $\frac{d}{2}$. The same is for Algorithm 6 versus d for Algorithm 5.

5. Lower Bounds

In this section we establish lower bounds on the competitive ratio in the WiFi and F2F models. Using bike sharing, first we prove a tight bound on the evacuation time when the robots know in which direction from the origin the exit is. Note that Theorem 7 can readily be used to compute the competitive ratio of the algorithms presented in Sections 3 and 4.

Theorem 7. *If the direction of the exit is known to the robots then evacuation time is* $\frac{v+1}{2v} \cdot d$ *and this is optimal.*

Proof of Theorem 7. Consider the algorithm whereby robot R_1 rides the bike for a distance x, releases the bike at x and continues by walking the remaining distance $d - x$, while robot R_2 walks for a distance x, picks up the bike at x and rides it for the remaining distance $d - x$. Note that robot R_1 reaches the exit at time $\frac{x}{v} + d - x$, while robot R_2 reaches the exit at time $x + \frac{d-x}{v}$. For the two robots to arrive at the same time it is required that $\frac{x}{v} + d - x = x + \frac{d-x}{v}$, which solves for $x = \frac{d}{2}$. Hence, the algorithm ensures that the two robots evacuate in time $\frac{1}{2} + \frac{1}{2v} = \frac{v+1}{2v}$.

Next we prove that the evacuation time above is optimal. If the robots never share the bike then the evacuation time will be d, which is also the arrival time of the slowest robot. So we may assume the robots share the bike. Let t_i be the termination time for robot

R_i in an optimal algorithm. Let x_i be the distance that robot R_i rides the bike. Without loss of generality let R_1 be the robot that fetches the bike from the origin. Clearly, this takes time $\frac{x_1}{v} + d - x_1$. Therefore $t_1 \geq \frac{x_1}{v} + d - x_1$. Similarly, for robot R_2 we have that $t_2 \geq \frac{x_2}{v} + d - x_2$. It follows that

$$
\begin{aligned}
\max\{t_1, t_2\} &\geq \frac{1}{2}\left(\frac{x_1}{v} + d - x_1 + \frac{x_2}{v} + d - x_2\right) \\
&= d - \frac{x_1}{2} - \frac{x_2}{2} + \frac{x_1}{2v} + \frac{x_1}{2v} \\
&= d - (x_1 + x_2)\frac{v-1}{2v} \\
&\geq d - \frac{v-1}{2v} \cdot d = \frac{v+1}{2v} \cdot d,
\end{aligned}
$$

where in the last inequality we used the fact that $x_1 + x_2 \leq d$, since by assumption only one robot can ride the bike at a time. This completes the proof of Theorem 7. $\square$

Using Theorem 7 we can prove the following result.

Theorem 8. *The evacuation time of any algorithm in either the WiFi or F2F model is bounded from below by* $\min\left\{\frac{d}{v} + \frac{v+1}{2v} \cdot d, d + \frac{v+1}{v} \cdot d\right\}$.

Proof of Theorem 8. Assume the two robots are starting at the origin and that the exit is placed at one of the two locations $\pm d$ which are unknown to the robots. Lets call the points $\pm d$ endpoints of the interval $[-d, +d]$. Without loss of generality assume that $-d$ is the first endpoint visited by a robot. There are two cases to consider depending on who is visiting this endpoint first.

Case 1: The biker visits $-d$ first.

To arrive at the endpoint $-d$ the biker has already spent time at least $\frac{d}{v}$ since he travels with speed v. At the time the biker arrives at $-d$ the hiker may be located either in the subinterval $[-d, 0]$ or in the subinterval $[0, +d]$. There are two subcases to consider depending on which of the two subintervals the hiker is located

- If the hiker is located in the interval $[-d, 0]$ then the adversary places the exit at $+d$ in which case by Theorem 7 the evacuation time will be at least $\frac{d}{v} + \frac{v+1}{2v} \cdot d$.
- If the hiker is located in the interval $[0, +d]$ then the adversary places the exit at $-d$ in which case again by Theorem 7 the evacuation time will be at least $\frac{d}{v} + \frac{v+1}{2v} \cdot d$.

Therefore regardless of the position of the hiker the evacuation time in this case is at least $\frac{d}{v} + \frac{v+1}{2v} \cdot d$.

Case 2: The hiker visits $-d$ first.

To arrive at the endpoint $-d$ the hiker has already spent time d since he travels with speed v. As before, the adversary will place the exit at the other endpoint $+d$. Unaided from the bike the hiker will take additional time $2d$ to arrive at the exit which is located at $+d$. However, in the wireless model the hiker can announce the exit has been found and therefore the two robots can share the bike to arrive at the exit $+d$. In view of Theorem 7 this takes additional time at least $\frac{v+1}{2v} \cdot 2d = \frac{v+1}{v} \cdot d$. Hence, in this case a lower bound on the evacuation time is $d + \frac{v+1}{v} \cdot d$.

Combining the two cases completes the proof of Theorem 8. $\square$

6. Conclusions

We proposed several evacuation algorithms in the wireless and face-to-face models. For each communication model we presented three algorithms which take advantage of the existence of the bike, a limited resource which can increase the search speed of the system of two robots. The resulting trajectories of the robots are specially designed so as to share the bike and ultimately reduce the overall evacuation time. We also discussed lower bounds.

The problem investigated is of theoretical nature and helps illuminate the trade offs between communication, and search time in search with mobile agents. Our study gives rise to several challenging open problems. For two robots, one could consider the problem when the speed of the bike depends on the hiker riding it. The case of multiple hikers and multiple bikes (not necessarily the same number) has never been investigated before. Additionally, one could also consider the case of faulty robots (crash or Byzantine). It would also be interesting to investigate other search domains such as stars and cycles or robots with reduced and/or enhanced capabilities.

Author Contributions: Conceptualization, K.J. and E.K.; methodology, K.J. and E.K.; resources, K.J. and E.K.; data curation, K.J. and E.K.; writing—original draft preparation, K.J. and E.K.; writing—review and editing, K.J. and E.K.; supervision, K.J. and E.K.; project administration, K.J. and E.K.; funding acquisition, K.J. and E.K. All authors have read and agreed to the published version of the manuscript.

Funding: Research of the second author was supported in part by NSERC Discovery grant.

Institutional Review Board Statement: Not applicanle.

Informed Consent Statement: Not applicanle.

Data Availability Statement: Data sharing not applicable.

Conflicts of Interest: The authors declare no conflict of interest.

References

1. Baeza Yates, R.; Culberson, J.; Rawlins, G. Searching in the plane. *Inf. Comput.* **1993**, *106*, 234–252. [CrossRef]
2. Beck, A. On the linear search problem. *Israel J. Math.* **1964**, *2*, 221–228. [CrossRef]
3. Bellman, R. An optimal search. *SIAM Rev.* **1963**, *5*, 274. [CrossRef]
4. Kao, M.Y.; Reif, J.H.; Tate, S.R. Searching in an unknown environment: An optimal randomized algorithm for the cow-path problem. *Inf. Comput.* **1996**, *131*, 63–79. [CrossRef]
5. Czyzowicz, J.; Kranakis, E.; Krizanc, D.; Narayanan, L.; Opatrny, J. Search on a Line with Faulty Robots. In Proceedings of the 2016 ACM Symposium on Principles of Distributed Computing, Chicago, IL, USA, 25–28 July, 2016; pp. 405–414.
6. Czyzowicz, J.; Georgiou, K.; Kranakis, E.; Krizanc, D.; Narayanan, L.; Opatrny, J.; Shende, S. Search on a Line by Byzantine Robots. *arXiv* **2016**, arXiv:1611.08209.
7. Czyzowicz, J.; Gasieniec, L.; Gorry, T.; Kranakis, E.; Martin, R.; Pajak, D. Evacuating Robots via Unknown Exit in a Disk. In Proceedings of the International Symposium on Distributed Computing, Austin, TX, USA, 12–15 October 2014; Springer: Berlin/Heidelberg, Germany, 2014; pp. 122–136.
8. Czyzowicz, J.; Georgiou, K.; Kranakis, E. Group Search and Evacuation. In *Distributed Computing by Mobile Entities, Current Research in Moving and Computing*; Springer: Cham, Switzerland, 2019; Volume 11340, pp. 335–370.
9. Kranakis, E.; Santoro, N.; Sawchuk, C.; Krizanc, D. Mobile agent rendezvous in a ring. In Proceedings of the IEEE 23rd International Conference on Distributed Computing Systems, Providence, RI, USA, 19–22 May 2003; pp. 592–599.
10. Czyzowicz, J.; Dobrev, S.; Kranakis, E.; Krizanc, D. The power of tokens: Rendezvous and symmetry detection for two mobile agents in a ring. In Proceedings of the International Conference on Current Trends in Theory and Practice of Computer Science, Nový Smokovec, Slovakia, 27–30 January 2019; Springer: Berlin/Heidelberg, Germany, 2008; pp. 234–246.
11. Czyzowicz, J.; Dobrev, S.; Godon, M.; Kranakis, E.; Sakai, T.; Urrutia, J. Searching for a Non-adversarial, Uncooperative Agent on a Cycle. *Theor. Comput. Sci.* **2020**, *806*, 531–542. [CrossRef]
12. Gasieniec, L.; Kijima, S.; Min, J. Searching with increasing speeds. In Proceedings of the International Symposium on Stabilizing, Safety, and Security of Distributed Systems, Tokyo, Japan, 4–7 November 2018, ; Springer: Berlin/Heidelberg, Germany, 2018; pp. 126–138.
13. Chrobak, M.; Gasieniec, L.; Gorry, T.; Martin, R. Group search on the line. In Proceedings of the International Conference on Current Trends in Theory and Practice of Informatics, Harrachov, Czech Republic, 27 July 2015; Springer: Berlin/Heidelberg, Germany, 2015; pp. 164–176.

14. Bampas, E.; Czyzowicz, J.; Gasieniec, L.; Ilcinkas, D.; Klasing, R.; Kociumaka, T.; Pajak, D. Linear search by a pair of distinct-speed robots. *Algorithmica* **2019**, *81*, 317–342. [CrossRef] [PubMed]
15. Jawhar, K. Bike Assisted Linear Search and Evacuation. Master's Thesis, School of Computer Science, Carleton University, Ottawa, ON, Canada, 2020.
16. Jawhar, K.; Kranakis, E. Bike Assisted Evacuation on a Line. In Proceedings of the SOFSEM (47th International Conference on Current Trends in Theory and Practice of Computer Science), Bozen-Bolzano, Italy, 25–28 January 2021; Springer: Berlin/Heidelberg, Germany, 2021.
17. Shaheen, S.A.; Guzman, S.; Zhang, H. Bikesharing in Europe, the Americas, and Asia: Past, present, and future. *Transp. Res. Rec.* **2010**, *2143*, 159–167. [CrossRef]
18. Chen, B.; Pinelli, F.; Sinn, M.; Botea, A.; Calabrese, F. Uncertainty in urban mobility: Predicting waiting times for shared bicycles and parking lots. In Proceedings of the 16th International IEEE Conference on Intelligent Transportation Systems (ITSC 2013), The Hague, The Netherlands, 6–9 October 2013; pp. 53–58.
19. Li, Z.; Zhang, J.; Gan, J.; Lu, P.; Gao, Z.; Kong, W. Large-scale trip planning for bike-sharing systems. *Pervasive Mob. Comput.* **2019**, *54*, 16–28. [CrossRef]
20. O'Mahony, E.; Shmoys, D.B. Data analysis and optimization for (citi) bike sharing. In Proceedings of the Twenty-Ninth AAAI Conference on Artificial Intelligence, Austin, TX, USA, 25–30 January 2015.
21. Pfrommer, J.; Warrington, J.; Schildbach, G.; Morari, M. Dynamic vehicle redistribution and online price incentives in shared mobility systems. *IEEE Trans. Intell. Transp. Syst.* **2014**, *15*, 1567–1578. [CrossRef]
22. Singla, A.; Santoni, M.; Bartók, G.; Mukerji, P.; Meenen, M.; Krause, A. Incentivizing users for balancing bike sharing systems. In Proceedings of the Twenty-Ninth AAAI Conference on Artificial Intelligence, Austin, TX, USA, 25–30 January 2015.
23. Yang, Z.; Hu, J.; Shu, Y.; Cheng, P.; Chen, J.; Moscibroda, T. Mobility modeling and prediction in bike-sharing systems. In Proceedings of the 14th Annual International Conference on Mobile Systems, Applications, and Services, Singapore, 26–30 June 2016; pp. 165–178.
24. Czyzowicz, J.; Georgiou, K.; Killick, R.; Kranakis, E.; Krizanc, D.; Narayanan, L.; Opatrny, J.; Pankratov, D. The Bike Sharing Problem. *arXiv* **2020**, arXiv:2006.13241.

Article

A Multi-Objective Optimization Problem on Evacuating 2 Robots from the Disk in the Face-to-Face Model; Trade-Offs between Worst-Case and Average-Case Analysis [†]

Huda Chuangpishit, Konstantinos Georgiou * and Preeti Sharma

Department of Mathematics, Ryerson University, 350 Victoria St., Toronto, ON M5B 2K3, Canada;
h.chuang@ryerson.ca (H.C.); preeti.sharma@ryerson.ca (P.S.)
* Correspondence: konstantinos@ryerson.ca
† This paper is an extended version of our paper published in An extended abstract of this work appeared in the
  proceedings of the 14th International Symposium on Algorithms and Experiments for Wireless Sensor
  Networks (ALGOSENSORS'18), Helsinki, Finland, 23–24 August 2018.

Received: 7 September 2020; Accepted: 26 October 2020; Published: 29 October 2020

Abstract: The problem of evacuating two robots from the disk in the face-to-face model was first introduced by Czyzowicz et al. [DISC'2014], and has been extensively studied (along with many variations) ever since with respect to worst-case analysis. We initiate the study of the same problem with respect to average-case analysis, which is also equivalent to designing randomized algorithms for the problem. In particular, we introduce constrained optimization problem $2\mathrm{EVAC}_{F2F}$, in which one is trying to minimize the average-case cost of the evacuation algorithm given that the worst-case cost does not exceed w. The problem is of special interest with respect to practical applications, since a common objective in search-and-rescue operations is to minimize the average completion time, given that a certain worst-case threshold is not exceeded, e.g., for safety or limited energy reasons. Our main contribution is the design and analysis of families of new evacuation parameterized algorithms which can solve $2\mathrm{EVAC}_{F2F}$, for every w for which the problem is feasible. Notably, the worst-case analysis of the problem, since its introduction, has been relying on technical numerical, computer-assisted calculations, following tedious robot trajectory analysis. Part of our contribution is a novel systematic procedure, which given *any evacuation algorithm*, can derive its worst- and average-case performance in a clean and unified way.

Keywords: evacuation; disk; face-to-face model; average-case analysis

---

1. Introduction

In search-type problems, several searchers (robots or mobile agents) are moving within a domain with the objective to identify the location of a (hidden) item. Several variations have been introduced for the problem that range among others with respect to the domain, to searcher specs, to the termination criterion (definition of feasibility) and to the objective of the underlying optimization problem. When in particular there are more than one searchers, and the termination criterion is that *all* searchers reach the hidden item (also known as exit), then the search problem is usually referred to as an *evacuation-type problem*. The term was introduced recently by Czyzowicz et al. in [1] who studied the problem of locating a hidden item on the perimeter of the unit disk with at least two robots. Once all searchers reach the hidden item (hence the name of the problem), the cost of the solution was defined as the time that the last searcher reaches the item (The termination criterion should not be confused with how the cost is quantified.). The problem was studied extensively also by subsequent publications exclusively under the lens of worst-case (competitive) performance, i.e., with the objective to minimize the worst-case (over all placements of the hidden item) time that the last searcher reaches the exit. In contrast, real-life applications are usually concerned with minimizing the average-case performance of an algorithm, especially if one must solve several instances of the problem. In our case, evacuation-type problems resemble search-and-rescue operations where on one hand it is desirable to minimize the average rescue time, but on the other hand one has to set hard thresholds regarding the possible worst-case performance of the operations (for example, due to safety concerns of the missing person/hidden item).

As a result, we are motivated to revisit existing evacuation-type problems in the realm of average-case performance, where the feasibility criterion is altered so as to also comply with possible bounds on the worst-case performance. Alternatively, one may think of an evacuation-type problem (or even search-type problem) as the multi-objective problem of minimizing simultaneously the worst-case and the average-case cost of feasible evacuation algorithm. In this direction, we study the evacuation problem [1] under the lens of average-case analysis with hard constraints on the worst-case performance, i.e., we study average-case worst-case trade-offs for the underlying optimization problem. More specifically, we introduce new evacuation-type problem $_2\mathrm{EVAC}_{F2F}^w$ with two searchers, in which the objective is to optimize the average-case evacuation time (or equivalently to minimize the performance of a randomized algorithm with unbounded access to randomness) condition on that the worst-case evacuation time does not exceed value w. The latter optimization problem seems to be particularly challenging in light of all previous results for the problem that relied on tedious calculations and/or on arguments tailored to the proposed algorithmic solutions. In that direction, we rely on a simple, novel and systematic theoretical approach that allows us to analyze the performance (both average-case and worst-case) of any evacuation algorithm based on computer-assisted calculations. Taking advantage of this approach, we introduce new families of algorithms for $_2\mathrm{EVAC}_{F2F}^w$. Interestingly, their worst-case analysis can be done rigorously, giving this way feasible evacuation protocols for the entire spectrum of values w. To motivate our results further, we also investigate the performance of existing algorithms of $_2\mathrm{EVAC}_{F2F}^w$, and we actually show that our new algorithms indeed outperform them for many values of w.

1.1. Related Work

In search problems, robots (or commonly referred to as mobile agents or searchers) are equipped with the task of efficiently locating a hidden item in a geometric domain. Studied in many variations since the 1960s, see seminal results [2,3], search problems first focused on identifying optimal probabilistic search and hiding strategies. Several subsequent studies of search-variations gave rise to numerous publications, and eventually to surveys, e.g., [4,5], and books [6–9] that also coined the umbrella term Search Theory. Over the years, search problems were also studied under the lens (i) of exploration by single [10–13] or multiple robots [14–16], (ii) of terrain mapping [17–19] and (iii) of hide-and-seek and pursuit-evasion objectives, e.g., see [20–23].

Although the first reference to a theoretical evacuation problem seems to date back to [24,25], the problem we study follows the line of research of Czyzowicz et al. [1]. Among the many results reported in [1], the one relevant to the current paper pertains to searching with two unit-speed robots for an item/exit hidden at the perimeter of a unit circle. The two searchers operate in the so-called face-to-face model that does not allow exchange of information from distance, and their goal is to minimize the worst-case evacuation time, i.e., the worst-case (over all placements of the exit) time that the last searcher reaches the exit. The first reported upper bound of 5.73906 [1] used a basic evacuation algorithm, according to which robots choose an arbitrary point on the circle, search in opposite directions, while the exit finder detours once the exit is found in order to catch and notify her peer before the evacuate together from the exit. By analyzing the worst placement of the exit [26], improved the upper bound to 5.628 by introducing a carefully chosen detour, before even the exit was reported, that was meant to expedite the catching phase in case the exit was found in a critical time window. The detour trajectory was later simplified and coupled with an elegant performance analysis that resulted in a further improvement of 5.625 [27]. Only recently, [28] reported yet another improvement of 5.6234, which was achieved by employing multiple detours. In contrast, the best lower bound known of 5.255 is due to [26], and no improvement has been reported since.

Since the inception of $_2$EVAC$_{F2F}$ in [1], numerous search and evacuation-type variants have emerged that studied different robot specs and/or number of searchers, different communication models, and different domains. Some notable examples include search and/or evacuation in the disk with more than 1 exits [29,30], in triangles [31,32], on multiple rays [33], in graphs [34,35], on a line with at least two robots [36,37] (generalizing the seminal result of [38]), with faulty robots [39–41] or with probabilistically faulty robots [42], with advice (information) [43], with priority specification on the searchers [44,45], with immobile agents [46,47], with time/energy trade-off requirements [48,49], with speed bounds [50], and with terrain dependent speeds [51], just to name some. The interested reader may also see the recent survey [52] that elaborates more on some selected topics.

In a different direction, Baeza-Yates et al. posed the question of minimizing the worst-case trajectory of a single robot searching for a target point at an unknown location in the plane [38]. This was generalized to multiple robots in [53], and later has been studied in [54,55]. More recently, Refs. [56,57] gave lower bounds for related problems. However, in these papers, the robots cannot communicate, and moreover, the objective is for the first robot to find the target.

1.2. Discussion on Closely Related Literature and Improvements

The problem of evacuating robots from the disc was first introduced in [1]. Among the many results reported in the latter paper, the one we expand upon is that of evacuating 2 robots in the face-to-face model. The algorithmic analysis was done exclusively in the deterministic model, and hence was worst-case. The reported upper bound of 5.628 was achieved by a simple algorithm, call it $\mathcal{A}$, that deploys both robots in an arbitrary point on the perimeter of the unit disc, and then makes them search in different directions. When the exit is found, the finder follows the shortest trajectory (straight line) in order to catch the non-finder in her trajectory so that together they return to the exit.

The first improvement of [26] was motivated by the monotonicity of the cost of $\mathcal{A}$ as a function of the time that the exit is found. A careful analysis shows that the cost function is concave, with the maximizer corresponding to the placement of exit that induces the worst-case performance. The weakness of the underlying communication model is that the (potential) non-finder, call it R_1 has no way to know whether her peer, R_2 has found the exit. At the same time, if enough time has passed and if R_1 has not been caught (notified) by R_2, then R_1 can deduce that her peer might have found the exit in the "dangerous" neighborhood of the placement inducing the worst-case cost, and hence R_2 might be already moving toward R_1 in order to notify her. As a result, R_1 has the incentive to deviate from searching the perimeter and to move toward the interior to expedite the (possible) rendezvous. Such a detour needs to be carefully chosen so that it is long enough to save cost in case the exit was indeed found in the dangerous neighborhood and short enough so that the detour does not add much in the cost in case it did not. Overall, the time that the detour occurs together with the direction and length of the detour give three degrees of freedom with an objective to optimize the cost in two different cases; when the exit is found in the "dangerous neighborhood" and when it is not. Call this (family of) algorithm(s) $\mathcal{A}_0$. The search space for choosing the three optimal parameters for $\mathcal{A}_0$ is indeed enormous, and at a high level [26] considered only a restriction of the previous idea in which robots agree on a forced meeting on the diameter of the circle, effectively reducing the degrees of freedom of the algorithm but simplifying the already technical and theoretical analysis of its worst-case cost, which eventually relied on numerical calculations.

The upper bound was later improved to 5.625 in [27]. The novelty of the latter work pertained to the introduction of an elegant theoretical analysis of algorithms $\mathcal{A}_0$, which on one hand avoided the forced meeting and on the other improved upon the bound of [26] by choosing optimally its three parameters, which were eventually computed by numerical calculations. The authors of [27] further mentioned that their results could potentially be improved by introducing, at a high level, a recursively defined sequence of detour points aiming to address the reduction of the optimal worst-case cost, should the detour points were one less (so the improvement of $\mathcal{A}$ to $\mathcal{A}_0$ can be thought as the basic step in the inductive argument). Indeed, ref. [28] used exactly that idea and reported a new upper bound of 5.6234.

Quite interestingly, all previous results pertaining to the worst-case analysis relied on arguments that were somehow tailored to the aforementioned algorithms (robot trajectories). In contrast, we provide a framework that allows us to compute, with computer-assisted calculations, the worst-case cost of any evacuation algorithm, as long as robot trajectories are described by unit-speed-inducing parameterized curves. It is the same framework that allows us also to also compute the average-case cost for any such algorithm.

The main improvement upon the previously results of the same problem pertains to minimizing the average-case cost. In one extreme, one can analyze the naive algorithm in which searchers always stay together, inducing very low average-case cost, but very high worst-case cost. In the other extreme, the optimal algorithm $\mathcal{A}_0$ of [27] has very low worst-case cost (nearly the best known) and, according to our results, relatively high average-case cost. As a result, we are motivated to minimize the average-case cost, subject to the worst-case cost ranges between the cost of the naive algorithm that of the optimal $\mathcal{A}_0$ (or equivalently to consider the multi-objective optimization problem of minimizing simultaneously the average-case cost and worst-case cost). Quite surprisingly, we verify, using our technical framework, that among the family of algorithms $\mathcal{A}_0$, the one inducing the minimum average-case cost is that of $\mathcal{A}$. It is the latter observation that gives the main motivation for introducing the three new parameterized families of evacuation algorithms that attempt to minimize their average-case cost, inducing worst-case cost that range, continuously, over the two extreme worst-case costs.

1.3. Outline of Our Results and Paper Organization

We introduce and study a multi-objective analog of the well-studied problem of evacuating two robots from the disk, performing in the face-to-face communication mode. More specifically, we introduce evacuation problem 2EVAC_{F2F}^{w} in which the two searchers need to evacuate in worst-case time no more than w, while still minimizing their average-case performance. One of our main contributions is a systematic method that allows us to perform both worst-case and average-case analysis for any evacuation algorithm that admits an analytic description. We apply our method to verify that previously known algorithms are not efficient (or not feasible) to 2EVAC_{F2F}^{w}, for certain values of w. As a result, we also motivate the introduction of three new families of evacuation algorithms that are feasible to 2EVAC_{F2F}^{w} for a wide spectrum of values w, and that induce a continuous bound for the efficient (pareto) frontier for the underlying multi-objective optimization problem. For our results we employ a rigorous analysis for the worst-case performance of the new algorithms, and we rely on numerical computer-assisted calculations and on the aforementioned novel systematic method for estimating their average-case performance.

The formal definition of our problem, along with a high-level exposition of our results appears in Section 2.1. In Section 2.2 our cornerstone observation pertaining to a systematic process for computing the average-case (and worst-case) performance of any evacuation algorithm that admits a convenient representation (as described in Section 2.3). Then in Section 3 we analyze two basic (benchmark) algorithms for 2EVAC_{F2F}^{w}. This allows us to motivate our problem for a range of w values that are related to the benchmark algorithms. In the same section, we also show that the families of previously proposed evacuation algorithms fail to be efficient for 2EVAC_{F2F}^{w}. In Section 4 we present new families of evacuation algorithms, which is also one of our main contributions. For these algorithms we give a rigorous worst-case performance analysis in Section 5, and computer-assisted average-case performance analysis in Section 6, based upon our results in Section 2.2. The reader can also find in the same section a formal quantification of our results for $_2\text{EVAC}^{w}_{F2F}$. Lastly, in Section 7 we conclude by discussing our findings and propose some future directions.

2. Preliminaries

2.1. Problem Definition and Main Results

The geometric domain of $_2\text{EVAC}_{F2F}$ is a unit disk, centered at the origin of the Cartesian plane. Two unit-speed searchers, starting from the origin, can move anywhere in the disk, and aim to identify a hidden object (exit) located at its perimeter (boundary). The object can be identified only by co-located robot, i.e., a robot that passes over it.

The two robots search in parallel, and perform under a centralized setting, i.e., they know each other's trajectories under the assumption that no exit is found. The underlying communication model is the so-called face-to-face that does not allow robots to exchange messages from distance, rather only when they meet. A *feasible evacuation* algorithm is a pair of trajectories, one for each robot, in which for every placement of the exit, both robots reach it eventually. For a technical reason, but also without loss of generality, we also require that robots will eventually stay at the exit idle. Placements of the exit i.e., instances to our problem, will be identified by $\text{cycle}(x) := (\cos(x), \sin(x))$, where $x \in [0, 2\pi)$. The evacuation time (cost) $\mathcal{C}(x)$ of a feasible evacuation algorithm on instance $\text{cycle}(x)$ is defined as the first time until both robots reach the exit. t.

In this work we are concerned with determining trade-offs between the worst-case and the average-case performance (of uniform placements of the exit) of evacuation algorithms for $_2\text{EVAC}_{F2F}$. More specifically, we say that an evacuation algorithm $\mathscr{A}$ with evacuation cost $\mathcal{C}(x)$ on instance $\text{cycle}(x)$ is (a, w)-*efficient* if

$$\text{Avg}\,(\mathscr{A}) := \text{E}_{x \in [0, 2\pi)}[\mathcal{C}(x)] \leq a,$$
$$\text{Wrs}\,(\mathscr{A}) := \sup_{x \in [0, 2\pi)}\{\mathcal{C}(x)\} \leq w.$$

where the expectation is with respect to the uniform distribution over $[0, 2\pi)$. Special to our problem is that $\text{Avg}\,(\mathscr{A})$ can also be interpreted as the expected performance of a randomized algorithm based on $\mathscr{A}$. Indeed, consider an algorithm which first performs a random rotation of the disk around the origin of angle θ, where θ is chosen uniformly at random from $[0, 2\pi)$, and then simulates $\mathscr{A}$. Please note that theoretically, this step requires infinitely many random bits, but one can simulate with any precision by enough many random bits. This random step is equivalent to choosing a deployment point uniformly at random on the disk. Due to the symmetry of the domain, it is irrelevant where the adversary will place the unique exit, and hence the expected performance of this randomized algorithm equals $\text{Avg}\,(\mathscr{A})$.

For algorithms $\mathscr{A}(p)$ parameterized by parameter(s) p, the pair $(\text{Avg}\,(\mathscr{A}(p)), \text{Wrs}\,(\mathscr{A}(p)))$ will correspond to a subset of $\mathbb{R}^2$ (and a curve if p is only one parameter) that we will refer to as the *Efficient Frontier*. We also adopt an optimization perspective of the problem, and we introduce the following optimization problem $_2\text{EVAC}^{w}_{F2F}$ on parameter w:

$$\min \frac{1}{2\pi} \int_0^{2\pi} \mathcal{C}(x)dx \qquad\qquad (_2\text{EVAC}^{w}_{F2F})$$
$$\text{s.t.}\quad \mathcal{C}(x) \leq w, \ \forall x \in [0, 2\pi).$$

Please note that the problem above is equivalent to the multi-objective optimization problem $\min\{\text{Avg}\,(\mathscr{A}(p)), \text{Wrs}\,(\mathscr{A}(p))\}$, and the conversion to the constrained problem above is due to the well-known ϵ-constraint method, e.g., see [58]. At the same time, no optimal lower bound is known for the worst-case cost in the case of unconstrained average-case cost, not to mention that lower bounds for similar problems are notoriously difficult. Consequently, under the current state-of-the-art it seems particularly challenging to determine the pareto frontier of the multi-objective optimization problem. Nevertheless, our arguments directly imply bounds for the pareto frontier.

As we show later, the values of w that make $_2\mathrm{EVAC}^w_{F2F}$ interesting lie between some special values w_1, w_2. Values w_1, w_2 are associated with benchmark algorithms, $\mathscr{B}_1, \mathscr{B}_2$, where in particular $\mathrm{Wrs}\,(\mathscr{B}_1) = w_1 \approx 5.739, \mathrm{Avg}\,(\mathscr{B}_1) = a_1 \approx 5.1172, \mathrm{Wrs}\,(\mathscr{B}_2) = w_2 \approx 7.283, \mathrm{Avg}\,(\mathscr{B}_1) = a_2 \approx 7.28319$. As a result, the reader may think of $\mathscr{B}_1$ as being inefficient in average case and efficient in worst case. Similarly, $\mathscr{B}_2$ is inefficient in worst case and efficient in average case.

As is the case for many variations of $_2\mathrm{EVAC}_{F2F}$, the cost of best solutions known are computed numerically. Our results pertain to upper bounds for a continuous spectrum of parameter w for problem $_2\mathrm{EVAC}^w_{F2F}$. In particular we propose families of algorithms $\mathscr{A}$ (over some parameters) so that as their parameters vary, we obtain $\mathrm{Wrs}\,(\mathscr{A}) = w$ and $\mathrm{Avg}\,(\mathscr{A}) = g(w)$, for each $w \in [w_1, w_2]$. The curve $(g(w), w)$ summarizing our results is depicted in Figure 1, and it is later quantified in Theorem 7 (see Section 5).

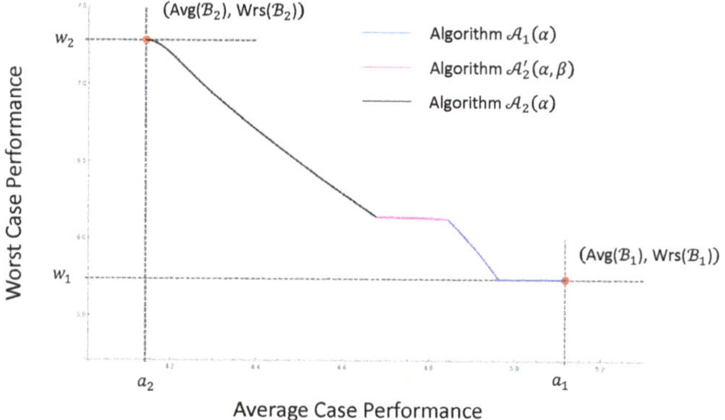

Figure 1. Illustration of the performance of our solution to $_2\mathrm{EVAC}^w_{F2F}$, for every $w \in [w_1, w_2]$. Depicted curve corresponds to parametric curve $(g(w), w)$, where $w, g(w)$ are the worst-case performance and average-case performance of three different families of evacuation algorithms $\mathscr{A}_1, \mathscr{A}_2', \mathscr{A}_2$, discussed formally in Section 4. Please note that the magenta curve is not a straight line and, as we show next, induces decreasing worst-case performance (as the average case performance increases).

Please note that an (a, w)-efficient algorithm gives a solution of value a for $_2\mathrm{EVAC}^w_{F2F}$. Our approach to prove Theorem 7 is to define families of evacuation algorithms $\mathscr{A}(p)$ parameterized by parameter(s) p. We will prove that these algorithms are $(u(p), v(p))$-efficient for some functions $u(p), v(p)$, and in particular the evaluation of the worst-case performance will be exact and monotone in p, while the computation of $v(p)$ will be computer-assisted. Then we will set $p = v^{-1}(w)$, and will be able to describe the average-case performance as a function of w as $g(w) := u(v^{-1}(w))$.

2.2. Computing Evacuation Times

For any feasible evacuation algorithm, we denote by $\mathcal{S}(x)$, the first time, after spending time 1 to reach the perimeter that $\mathrm{cycle}(x)$ is visited by any robot. In other words, $\mathcal{S}(x)$ is the time robots spend searching till the exit is found for the first time, assuming that robots do not waste time in the interior of the circle before they start searching. Clearly, when a robot, say $\mathscr{R}_1$, locates the exit at $\mathrm{cycle}(x)$, it may attempt to catch $\mathscr{R}_2$ while moving along $\mathscr{R}_2$'s trajectory along the shortest line segment, say of length $\mathcal{E}(x)$. Once robots meet, they return together to $\mathrm{cycle}(x)$, inducing total evacuation cost $\mathcal{C}(x) = 1 + \mathcal{S}(x) + 2\mathcal{E}(x)$.

All existing results for $_2$EVAC$_{F2F}$, from a worst-case complexity perspective, rely on numerical computer-assisted estimation of $\sup_x C(x)$, after identifying properties of the maximizer. In this section, we elevate existing arguments, and we propose a generalized and unified approach for computing $C(x)$, for any x and for any robot trajectories. For the sake of formality, as well as for practical purposes, robot trajectories will be defined by parametric functions $\mathcal{F}(t) = (f(t), g(t))$, where $f, g : \mathbb{R} \mapsto \mathbb{R}$ are continuous and piecewise differentiable. In particular, search protocols for the two robots will be given by trajectories $\mathscr{R}_1(t), \mathscr{R}_2(t)$, where $\mathscr{R}_i(t)$ will denote the position of robot $\mathscr{R}_i$ at time $t \geq 0$. Therefore, any evacuation algorithm will be identified by a tuple $(\mathscr{R}_1, \mathscr{R}_2)$. To simplify notation, we will only determine the trajectories from the moment the two robots reach the perimeter of the circle, and until the entire circle is searched (under the assumption that no exit is found), and we will silently assume that robots stay put after exploration is over.

Lemma 1. *Consider instance* cycle(x) *of* $_2$EVAC$_{F2F}$, *and suppose that for a feasible evacuation algorithm* $(\mathscr{R}_1, \mathscr{R}_2)$, *robot 1 is the first robot that finds the exit. Then* $\mathcal{E}(x) = \bar{t} - \mathcal{S}(x)$, *where* $\bar{t} = \bar{t}(x)$ *is the smallest root, no less than* $\mathcal{S}(x)$, *of function*

$$h_x(t) := \|\mathscr{R}_2(t) - \mathscr{R}_1(\mathcal{S}(x))\| - t + \mathcal{S}(x). \tag{1}$$

Proof. The reader may consult Figure 2 that complements our argument.

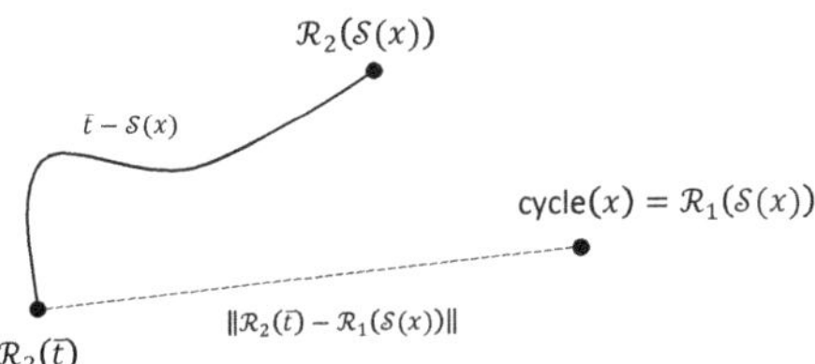

Figure 2. An abstract depiction of the trajectories of $\mathscr{R}_1, \mathscr{R}_2$, assuming that $\mathscr{R}_1$ is the finder of the exit, located at cycle(x) at time $\mathcal{S}(x)$. Time $\bar{t}$ is the time, after they start searching for the exit on the perimeter that the two robots meet at $\mathscr{R}_1(\bar{t}) = \mathscr{R}_2(\bar{t})$.

First observe that $h_x(t)$ is continuous. If the robots find the exit together then result holds trivially. So we may assume that that the two robots are not co-located when the exit is found, in which case we have $h_x(\mathcal{S}(x)) > 0$. At the same time, since the evacuation algorithm is feasible, $\mathscr{R}_2(t)$ is eventually a constant, and hence for big enough t we have that $h_x(t)$ becomes eventually negative. By the mean value theorem, there is $t_0 > 0$ for which $h_x(t_0) = 0$.

Now consider the smallest positive root $\bar{t}$ of h_x, no less than $\mathcal{S}(x)$. At time $\bar{t}$, $\mathscr{R}_2$ is located at point $\mathscr{R}_2(\bar{t})$, and it is $\|\mathscr{R}_2(\bar{t}) - \mathscr{R}_1(\mathcal{S}(x))\|$ away from the location cycle(x) of the discovered exit. At the same time, $\mathscr{R}_1$ moves with speed 1 along the shortest path to catch $\mathscr{R}_2$ in her trajectory. Hence it takes $\mathscr{R}_1$ some $\bar{t} - \mathcal{S}(x)$ extra time from the moment the exit is found until she reaches point $\mathscr{R}_2(\bar{t})$. By definition we have $\mathscr{R}_1(\bar{t}) = \mathscr{R}_2(\bar{t})$, and therefore $\mathcal{E}(x) = \bar{t} - \mathcal{S}(x)$ as claimed. $\square$

For some special trajectories, $\mathcal{E}(x)$ admits a simpler description that we describe next. Before that, we introduce some notation pertaining to a function $\delta : [0, \pi] \mapsto \mathbb{R}_+$, which we widely use in the remaining of the paper:

$$\delta(x) := \text{ unique non-negative root (regarding } d) \text{ of } \text{``} 2\sin\left(x + \frac{d}{2}\right) = d \text{''}. \tag{2}$$

To simplify notation, we will also abbreviate $\delta(x)$ by δ_x. To show that δ_x is well-defined, consider function $f_x(d) = 2\sin\left(x + \frac{d}{2}\right) - d$. The derivative of the function is $f'_x(d) = \cos x + \frac{d}{2} - 1 \leq 0$.

Since $f_x'(d)$ has also a unique root in $[0, \pi]$, it follows that $f_x(d)$ is strictly decreasing. Observe now that $f_x(0) = 2\sin(x) \geq 0$, while $f_x(2) = 2\sin(x+1) - 2 \leq 0$. We conclude that due to the strict monotonicity of $f_x(d)$, the latter function has indeed unique root in $d \in [0, 2]$. Finally, it is easy to see that $f_x(d) < 0$ when $d > 2$.

Lemma 2. *For some instance* $\text{cycle}(x)$ *of* $_2\text{EVAC}_{F2F}$, *suppose that for a feasible evacuation algorithm* $(\mathscr{R}_1, \mathscr{R}_2)$, $\mathscr{R}_1$ *is the finder of the exit, say at time* $t_0 = \mathcal{S}(x)$. *Assume that both* $\mathscr{R}_1(t_0), \mathscr{R}_2(t_0)$ *lie on the circle at arc distance* 2α, *and suppose that* $\mathscr{R}_2$'s *movement is along the perimeter of the circle toward the complementary arc of length* $2\pi - \alpha$. *Then,* $\mathcal{E}(x) = \delta_\alpha$.

Proof. The lemma follows by applying transformation $t - \mathcal{S}(x) = d$ in the definition of $h_x(t)$ in Lemma 1, so that $\mathcal{E}(x) = t - \mathcal{S}(x) = d$. $\square$

We are ready to conclude with a corollary that will be handy for computing evacuation times numerically, and without relying on excessive case analysis, as was the case before.

Corollary 1. *Consider feasible evacuation algorithm* $(\mathscr{R}_1, \mathscr{R}_2)$ *for* $_2\text{EVAC}_{F2F}$. *For any instance* $\text{cycle}(x)$ *for which* $\mathscr{R}_1$ *is the exit finder, the evacuation cost can be computed as* $C(x) = 1 + 2\bar{t} - \mathcal{S}(x)$, *where* $\bar{t} = \bar{t}(x)$ *is the smallest root, at least* $\mathcal{S}(x)$, *of* $h_x(t) := \|\mathscr{R}_2(t) - \mathscr{R}_1(\mathcal{S}(x))\| - t + \mathcal{S}(x)$.

2.3. Trajectory Description

Robot trajectories will be described in phases. We will always omit the "deployment phase", i.e., the movement from the circle center to its perimeter, and we will only describe the trajectories from the moment robots start searching the circle. In each phase, robot $\mathscr{R}$, will be moving between two explicit points, either along an arc, or along a line segment (chord of an arc), see Observations 1 and 2 below. We will summarize robot trajectories in tables of the following format.

Robot	Phase #	Trajectory	Duration
$\mathscr{R}$	1	$\mathscr{R}(t)$	t_1
	2	$\mathscr{R}(t)$	t_2
	$\vdots$		$\vdots$

To ease notation, trajectory $\mathscr{R}(t)$ of phase i will be described with parametric equations as if the time is reset to 0 after time $t_0 + t_1 + t_2 + \ldots + t_{i-1}$, where $t_0 = 1$ (this is the time that robots reach the circle). The two fundamental trajectory components are movements along arcs and movements along line segments.

Observation 1. *Let* $b \in [0, 2\pi)$ *and* $\sigma \in \{-1, 1\}$. *The trajectory of an object moving at speed 1 on the perimeter of a unit circle with initial location* $\text{cycle}(b)$ *is given by the parametric equation* $\text{cycle}(\sigma t + b) = (\cos(\sigma t + b), \sin(\sigma t + b))$. *If* $\sigma = 1$ *the movement is counterclockwise (ccw), and clockwise (cw) otherwise.*

Proof. It is immediate that when $t = 1$, the object is located at $\text{cycle}(b)$. Its speed is given by calculating $\left\|\frac{\partial}{\partial t}\text{cycle}(\sigma t + b)\right\|$. Indeed, we have

$$\left(\frac{d}{dt}\cos(\sigma t + b)\right)^2 + \left(\frac{d}{dt}\sin(\sigma t + b)\right)^2 = \sigma^2(-\sin(\sigma t + b))^2 + \sigma^2(\cos(\sigma t + b))^2 = 1,$$

as wanted. $\square$

Observation 2. *Consider distinct points $A = (a_1, a_2), B = (b_1, b_2)$ in $\mathbb{R}^2$. The trajectory of a speed 1 object moving along the line passing through A, B and with initial position A is given by the parametric equation*

$$\text{line}(A, B, t) := \left(\frac{b_1 - a_1}{\|A - B\|} t + a_1, \frac{b_2 - a_2}{\|A - B\|} t + a_2 \right).$$

Proof. By definition, the parametric equation above corresponds to a line. Elementary calculations show that $\text{line}(A, B, 0) = A$ and $\text{line}(A, B, \|A - B\|) = B$, that is the object starts at A, and that it passes through B. Object speed is calculated as $\left\| \frac{\partial}{\partial t} \text{line}(A, B, t) \right\|$. In that direction we have

$$\left(\frac{d}{dt} \left(\frac{b_1 - a_1}{\|A - B\|} t + a_1 \right) \right)^2 + \left(\frac{d}{dt} \left(\frac{b_2 - a_2}{\|A - B\|} t + a_2 \right) \right)^2 = \left(\frac{b_1 - a_1}{\|A - B\|} \right)^2 + \left(\frac{b_2 - a_2}{\|A - B\|} \right)^2 = 1$$

as promised. $\square$

Finally, the analysis of our algorithm trajectories will give rise to several constants. For the reader's convenience, we list here the numerical values of the most common constants that will be encountered later. $w_1 \approx 5.73906, w_0 \approx 6.11953, w' \approx 6.12851, w_2 \approx 7.28319, \alpha' \approx 1.15468, \bar{\alpha} \approx 1.54419, \beta' \approx 0.0241653, \beta_0 \approx 0.04388$.

All constants are formally defined when they are first introduced.

3. Two Benchmark Algorithms and Motivation

In this section, we describe two benchmark algorithms for $_2\text{EVAC}_{F2F}$, as well as perform average-case analysis to algorithms previously proposed in the literature. The reader may consult Figure 3 for the algorithms analyzed in this section.

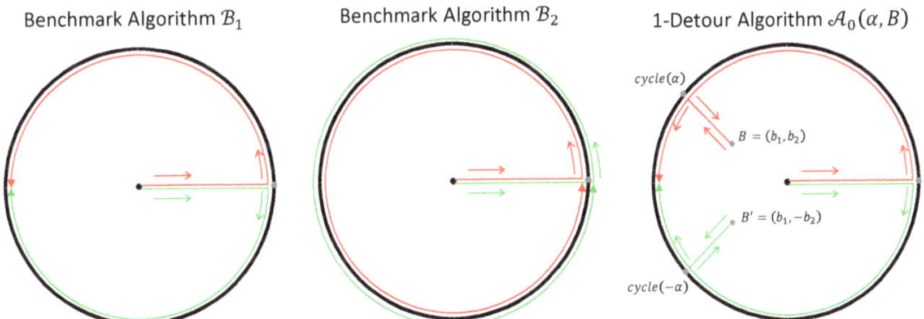

Figure 3. Robot Trajectories for algorithms $\mathcal{B}_1, \mathcal{B}_2, \mathcal{A}_0$. The depicted trajectories show the search of the circle, and not the evacuation step that is performed once the exit is found.

Czyzowicz et al. [1] were the first to introduce an evacuation algorithm for $_2\text{EVAC}_{F2F}$, which we denote here by $\mathcal{B}_1$ (see Figure 3 on the left).

Definition 1 (Benchmark Algorithm $\mathcal{B}_1$). *For all $t \in [0, \pi]$, $\mathcal{R}_1(t) = cycle(t)$ and $\mathcal{R}_2(t) = cycle(-t)$.*

Observation 3. *Benchmark Algorithm $\mathcal{B}_1$ is $(5.1172, 5.73906)$-efficient.*

Proof. Please note that it takes time π to search the entire circle, and that the two trajectories are symmetric with respect to horizontal axis. Therefore, we may assume that the instance $cycle(x)$ satisfies $x \in [0, \pi]$.

Clearly, for any such x, we have that $\mathcal{S}(x) = x$. By Lemma 2, we have that $\mathcal{C}(x) = 1 + \mathcal{S}(x) + 2\mathcal{E}(x) = 1 + x + 2\delta_x$. Numerical calculations (software assisted) show that

$$\text{Wrs}(\mathscr{B}_1) = \sup_{x \in [0,\pi]} \{\mathcal{C}(x)\} = \sup_{x \in [0,\pi]} \{1 + x + 2\delta_x\} \approx 5.73906,$$

$$\text{Avg}(\mathscr{B}_1) = \mathrm{E}_{x \in [0,\pi]}[\mathcal{C}(x)] = \frac{1}{\pi} \int_{x=0}^{\pi} (1 + x + 2\delta_x)\, dx \approx 5.1172.$$

$\square$

$\mathscr{B}_1$ should be understood as being efficient in the worst case, but inefficient on average. The claim becomes transparent by introducing the following *naive* algorithm for $_2\text{EVAC}_{F2F}$ that we depict in the middle of Figure 3.

Definition 2 (Benchmark Algorithm $\mathscr{B}_2$). *For each $t \in [0, 2\pi]$, $\mathscr{R}_1(t) = \mathscr{R}_2(t) = cycle(t)$.*

Observation 4. *Benchmark Algorithm $\mathscr{B}_2$ is $(1 + \pi, 1 + 2\pi)$-efficient.*

Proof. It is easy to see that for all $x \in [0, 2\pi)$ we have $\bar{f}(x) = \mathcal{S}(x) = x$ and $\mathcal{E}(x) = 0$. Therefore $\mathcal{C}(x) = 1 + x$, and hence

$$\text{Wrs}(\mathscr{B}_2) = \sup_{x \in [0,2\pi)} \{\mathcal{C}(x)\} = 1 + 2\pi,$$

$$\text{Avg}(\mathscr{B}_2) = \mathrm{E}_{x \in [0,2\pi)}[\mathcal{C}(x)] = \int_{x=0}^{2\pi} (1 + x)\, dx = 1 + \pi.$$

$\square$

$\mathscr{B}_2$ should be understood as highly efficient on average, but inefficient in the worst case. Moreover, it should be clear that $\mathscr{B}_1, \mathscr{B}_2$ are feasible solutions to $_2\text{EVAC}_{F2F}^w$, for $w = 5.1172$ and $w = 1 + 2\pi$, respectively. We conjecture that $\mathscr{B}_1$ is indeed the optimal evacuation algorithm among all algorithms with worst-case performance no more than $1 + 2\pi$. At the same time, below we show that $\mathscr{B}_2$ is the best algorithm for $_2\text{EVAC}_{F2F}^w$, when $w = 5.1172$, among those previously used to improve upon the worst-case performance up to the third decimal. The importance of this observation is two-fold; first we are motivated to study $_2\text{EVAC}_{F2F}^w$ for the entire spectrum of $w \in [\text{Wrs}(\mathscr{B}_1), \text{Wrs}(\mathscr{B}_2)]$, and second we deduce that in order to perform well on average, we need to devise and analyze new evacuation algorithms.

Upper bounds for the worst-case performance of $\mathscr{B}_1$ were later improved first to 5.628 [26], then to 5.625 [27], and then to 5.623 [28]. The main idea behind the improvement is to understand the monoticity of $\mathcal{C}(x)$ for algorithm $\mathscr{B}_1$. Indeed, the following lemma was implicit in both [26,27], and can be obtained numerically.

Lemma 3. *There is α_0, where $\alpha_0 \approx 0.96782$, so that evacuation cost $\mathcal{C}(x)$ of $\mathscr{B}_1$ for $_2\text{EVAC}_{F2F}$ on instance cycle(x) is strictly increasing for $x \in [0, \alpha_0]$, and strictly decreasing in $x \in [\alpha_0, \pi]$. In particular, $\text{Wrs}(\mathscr{B}_1) = \mathcal{C}(\alpha_0) \approx 5.73906$.*

Consider now an execution of $\mathscr{B}_1$ in which one of the robots, say $\mathscr{R}_2$ continues searching on the circle and is close to approach a location that would be the meeting point if the instance was cycle(α_0). In an attempt to help expedite a potential meeting (in case $\mathscr{R}_1$ is approaching) and effectively reducing the cost of the worst case, $\mathscr{R}_2$ would make a minor detour toward the interior of the disk, before returning back to the exploration of the circle. This simple idea was explored in [26] and in [27] where the following family of algorithms were introduced, parameterized by $\alpha \in [0, \pi]$ and point B within the unit disk, see also right of Figure 3 (the simplified version presented here is due to [27]).

Definition 3 (1-Detour Algorithm $\mathscr{A}_0(\alpha, B)$)**.** *For all $t \in [0, \pi + 2\,\|\mathrm{cycle}(\alpha) - B\|]$, the trajectory of $\mathscr{R}_1$ is defined as*

Robot	Phase #	Trajectory	Duration
$\mathscr{R}_1$	1	$\mathrm{cycle}(t)$	α
	2	$\mathrm{line}(\mathrm{cycle}(\alpha), B, t)$	$\|\mathrm{cycle}(\alpha) - B\|$
	3	$\mathrm{line}(B, \mathrm{cycle}(\alpha), t)$	$\|\mathrm{cycle}(\alpha) - B\|$
	4	$\mathrm{cycle}(t + \alpha)$	$\pi - \alpha$

The trajectory of $\mathscr{R}_2$ is symmetric with respect to the horizontal axis.

The crux of the contribution of [26] was to prove that there exists α, B for which the worst-case performance is no more than 5.644 (and a delicate refinement is needed to achieve 5.628). Notably, their analysis is tedious and lengthy, whereas we can obtain the same result, relying again on numerical calculations, with minimal effort. Then, [27] proposed variations of $\mathscr{A}_0(\alpha, B)$ in which each robot performs more than 1 detours (see Phases 2,3 of $\mathscr{A}_0(\alpha, B)$), giving rise to [28]. Hence, t-detour algorithms are parameterized by a sequence $\alpha_1, \ldots, \alpha_t$, where $\alpha_i \geq 0$ and $\sum_i \alpha_i \leq \pi$, and points B_i in the disk. Even 2-detour algorithms achieve worst-case performance 5.623, while for each $t \geq 2$, t-detour algorithms do induce strictly improved performance (for appropriate choices of the parameters) but the improvement seems to be negligible.

Motivated by the results in [26,27], one is tempted to ask whether any algorithm in the family $\mathscr{A}_0(\alpha, B)$ improves upon $\mathscr{B}_1$ with respect to the average-case analysis. The next claim is due to exhaustive, computer-assisted numerical calculations, see also Figure 4.

Theorem 5. *For every $\alpha \in [0, \pi)$ and for every B in the unit disk $\mathrm{Avg}\,(\mathscr{A}_0(\alpha, B)) \geq \mathrm{Avg}\,(\mathscr{B}_1)$.*

Theorem 5 provides strong motivation for studying problem $_2\mathrm{EVAC}^w_{\mathrm{F2F}}$, since it shows that in order to establish good upper bounds, i.e., our main results depicted in Figure 1 and quantified later in Theorem 7, one needs to employ new evacuation algorithms. Recall that even $\mathrm{Wrs}\,(\mathscr{B}_1)$ and $\mathrm{Wrs}\,(\mathscr{A}_0(\alpha, B))$ were estimated with computer-assisted calculations. Due to the nature of the problem, we are bound to rely on computer-assisted calculations as well. Notably, our much more intense computational work is feasible only because we employ the new method for computing evacuation times due to Corollary 1 and Definition 3 of $\mathscr{A}_0(\alpha, B)$ trajectories. Overall, to verify Theorem 5 we compute pairs $(\mathrm{Avg}\,(\mathscr{A}_0(\alpha, B))\,, \mathrm{Wrs}\,(\mathscr{A}_0(\alpha, B)))$ for more than 500,000 different parameter values and we depict them in Figure 4.

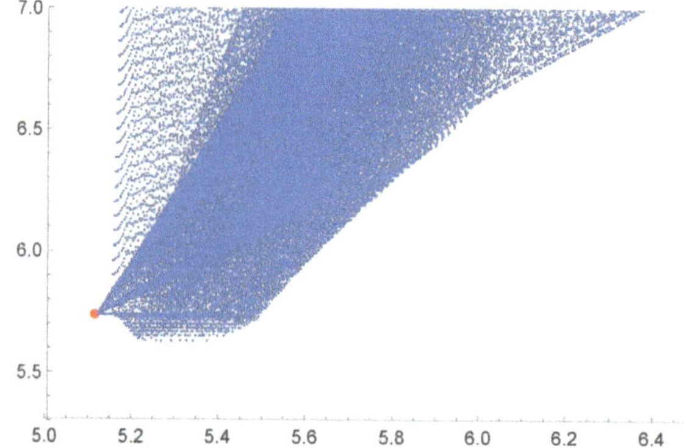

Figure 4. Performance analysis of $\mathscr{A}_0(\alpha, B)$ for various values of parameters α, B. Blue points (a, w) correspond to (a, w)-efficient algorithms $\mathscr{A}_0(\alpha, B)$. The red point is $(\mathrm{Avg}\,(\mathscr{B}_1), \mathrm{Wrs}\,(\mathscr{B}_1))$, i.e., the performance of $\mathscr{B}_1$ in the average-worst-case space. Please note that no algorithm $\mathscr{A}_0$ performs better on average than $\mathscr{B}_1$, while all $\mathscr{A}_0(t, \mathrm{cycle}(t))$ is exactly $\mathscr{B}_1$ for every point $t \in [0, \pi]$. Notably, all points lie above the threshold of worst-case performance 5.625, and some are arbitrarily close to that value (corresponding to choices of α, B that give the algorithm of [27]).

4. New Evacuation Algorithms

In this section, we propose families of evacuation algorithms for problem $_2\mathrm{EVAC}^w_{F2F}$, for the entire spectrum of $w \in [\mathrm{Wrs}\,(\mathscr{B}_1), \mathrm{Wrs}\,(\mathscr{B}_2)]$. Our algorithms are summarized in Figure 5.

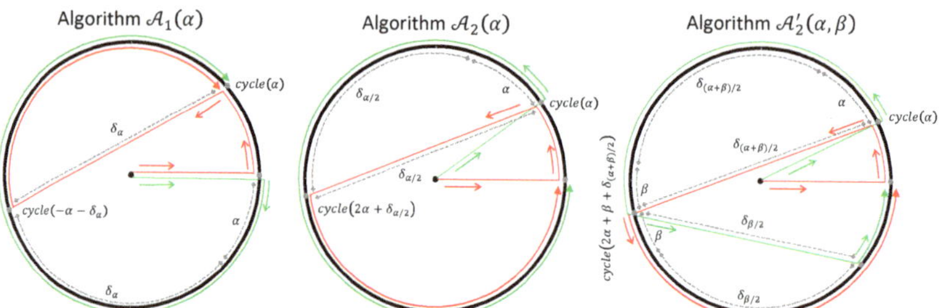

Figure 5. Robot Trajectories for algorithms $\mathscr{A}_1, \mathscr{A}_2, \mathscr{A}'_2$. The depicted trajectories show the search of the circle, and not the evacuation step that is performed once the exit is found. Arcs that are searched by both robots are also searched simultaneously, i.e., robots are co-located and search together.

First we define families of evacuation algorithms that, as we show next, perform well for $_2\mathrm{EVAC}^w_{F2F}$ in the "neighborhood of $\mathscr{B}_1$", i.e., for w close to $\mathrm{Wrs}\,(\mathscr{B}_1)$. Our algorithms are parameterized by α, and their circle exploration lasts $2\pi - \alpha$.

Definition 4 (Algorithm $\mathscr{A}_1(\alpha)$). *For all $t \in [0, 2\pi - \alpha]$, the trajectory of $\mathscr{R}_1$ is defined as*

Robot	Phase #	Trajectory	Duration
$\mathscr{R}_1$	1	$\mathrm{cycle}(t)$	α
	2	$\mathrm{line}(\mathrm{cycle}(\alpha), \mathrm{cycle}(-\alpha - \delta_\alpha), t)$	δ_α
	3	$\mathrm{cycle}(-\alpha - \delta_\alpha - t)$	$2\pi - 2\alpha - \delta_\alpha$

where δ_α is defined in (2). *The trajectory of $\mathscr{R}_2$ is defined as $\mathscr{R}_2(t) = \mathrm{cycle}(-t)$, for all $t \in [0, 2\pi - \alpha]$.*

$\mathscr{A}_1$ is depicted in Figure 5 on the left. At a high level $\mathscr{A}_1(\alpha)$ is a modification of $\mathscr{B}_1$ that is based on the following idea. The execution of $\mathscr{A}_1(\alpha)$ is the same as in $\mathscr{B}_1$ until each robot searches an arc of length α (and hence $\mathscr{A}(\pi)$ coincides with $\mathscr{B}_1$). After time α, $\mathscr{R}_1$ abandons her trajectory and catches $\mathscr{R}_2$, on the perimeter of the circle resembling a trajectory as if the exit was located at $\mathscr{R}_1(\alpha)$. It is not difficult to see that the definition of δ_α above satisfies $\mathscr{R}_1(\alpha + \delta_\alpha) = \mathscr{R}_2(\alpha + \delta_\alpha) = \text{cycle}(-\alpha - \delta_\alpha)$.

Next we define a family of algorithms $\mathscr{A}_2$ which, as we show later, perform well in the "neighborhood of $\mathscr{B}_2$", i.e., for w close to Wrs $(\mathscr{B}_2)$. For this recall definition (2) of δ_a. We let $\gamma_0 \approx 2.2412$ be the root of $2\alpha + \delta_{\alpha/2} = 2\pi$. For every $\alpha \leq \gamma_0$ we define a family of algorithms on parameter α whose circle exploration lasts $2\pi - \alpha$.

Definition 5 (Algorithm $\mathscr{A}_2(\alpha)$). *For all $t \in [0, 2\pi - \alpha]$, the trajectory of $\mathscr{R}_1$ is defined as*

Robot	Phase #	Trajectory	Duration
$\mathscr{R}_1$	1	$\text{cycle}(t)$	α
	2	$\text{line}(\text{cycle}(\alpha), \text{cycle}(2\alpha + \delta_{\alpha/2}), t)$	$\delta_{\alpha/2}$
	3	$\text{cycle}(2\alpha + \delta_{\alpha/2} + t)$	$2\pi - 2\alpha - \delta_{\alpha/2}$

The trajectory of $\mathscr{R}_2$ is defined as $\mathscr{R}_2(t) = \text{cycle}(\alpha + t)$, for all $t \in [0, 2\pi - \alpha]$.

$\mathscr{A}_2$ is depicted in the middle of Figure 5. The condition that $\alpha \leq \gamma_0$ is added for simplicity to ensure that the latest catching point occurs while the other robot is still searching, and is not mandatory. At a high level $\mathscr{A}_2(\alpha)$ is a generalization of $\mathscr{B}_2$ (note that $\mathscr{A}_2(0) = \mathscr{B}_2$). For the first α time units, robots search in the same direction until $\mathscr{R}_1$ arrives at the deployment point of $\mathscr{R}_2$. Then, $\mathscr{R}_1$ catches $\mathscr{R}_2$ on the circle, as if the exit was located at $\mathscr{R}_1(\alpha)$ (which by Lemma 2 happens in $\delta_{\alpha/2}$ extra time).

Finally, we introduce a family of evacuation algorithms which will perform well for $_2\text{EVAC}^w_{\text{F2F}}$ for intermediate values of $w \in [\text{Wrs}(\mathscr{B}_1), \text{Wrs}(\mathscr{B}_2)]$. For this we generalize family $\mathscr{A}_2$ so that the two robots perform two alternating jumps, with parameters α, β satisfying $2\alpha + 2\beta + \delta_{(\alpha+\beta)/2} + \delta_{\beta/2} \leq 2\pi$, see right of Figure 5.

Definition 6 (Algorithm $\mathscr{A}_2'(\alpha, \beta)$). *For notational convenience, we set $\zeta_{\alpha,\beta} := 2\alpha + \beta + \delta_{(\alpha+\beta)/2}$. For all $t \in [0, 2\pi - \alpha - \beta]$, the trajectories of $\mathscr{R}_1, \mathscr{R}_2$ are defined as follows*

Robot	Phase #	Trajectory	Duration
$\mathscr{R}_1$	1	$\text{cycle}(t)$	α
	2	$\text{line}(\text{cycle}(\alpha), \text{cycle}\left(\zeta_{\alpha,\beta}\right), t)$	$\delta_{(\alpha+\beta)/2}$
	3	$\text{cycle}\left(\zeta_{\alpha,\beta} + t\right)$	$2\pi - 2\alpha - \beta - \delta_{(\alpha+\beta)/2}$
$\mathscr{R}_2$	1	$\text{cycle}(\alpha + t)$	$\alpha + \beta + \delta_{(\alpha+\beta)/2}$
	2	$\text{line}(\text{cycle}\left(\zeta_{\alpha,\beta}\right), \text{cycle}\left(\zeta_{\alpha,\beta} + \delta_{\beta/2}\right), t)$	$\delta_{\beta/2}$
	3	$\text{cycle}\left(\zeta_{\alpha,\beta} + \beta + \delta_{\beta/2} + t\right)$	$2\pi - 2\alpha - 2\beta - \delta_{(\alpha+\beta)/2} - \delta_{\beta/2}$

Next we describe the meaning of parameters α, β of the Robot trajectories above. As in the family of algorithms $\mathscr{A}_2$, parameter α represents the arc distance the two robots have before the one preceding decides to jump ahead. In $\mathscr{A}_2$ the two robots meet again once the jumper reaches the perimeter of the circle. In $\mathscr{A}_2'$ the jumper deploys a little further away on the circle so that when the other robot reaches the deployment point of the jumper, the two robots are at arc distance β. As a result, the time it takes both robots to complete searching the entire circle is $2\pi - \alpha - \beta$, as well as $\mathscr{A}_2(\alpha, 0)$ coincides with $\mathscr{A}_2(\alpha)$. Finally, note that even though $\mathscr{A}_2'$ will be eventually invoked for seemingly restricted values of β ($\beta \leq \beta_0 \approx 0.04388$), the deviation in the performance will be significant enough (e.g., $\delta_{\beta_0/2} \approx 0.977997$) to account for its use in our upper bounds.

5. Worst-Case Performance Analysis

In this section, we perform worst-case analysis for all algorithmic families $\mathscr{A}_1, \mathscr{A}_2, \mathscr{A}_2'$ with respect to their parameters. Notably, results in this section are quantified formally and exactly by closed formulas. At a high level, each of $\mathscr{A}_1, \mathscr{A}_2, \mathscr{A}_2'$ will be invoked to solve $_2\text{EVAC}^w_{\text{F2F}}$ for different values of $w \in [\text{Wrs}(\mathscr{B}_1), \text{Wrs}(\mathscr{B}_2)]$, and each of them will have competitive average-case performance for the corresponding worst-case performance w. As an easy warm-up, we analyze $\mathscr{A}_1$.

Lemma 4 (Worst-Case Analysis for $\mathscr{A}_1$). *Let* $\bar{\alpha} = 1 + 2\pi - w_1$, *where* $w_1 = \text{Wrs}(\mathscr{B}_1)$. *Then, for all* $\alpha \in [0, \pi]$, *we have that*

$$\text{Wrs}(\mathscr{A}_1(\alpha)) = \begin{cases} 1 + 2\pi - \alpha & , \forall \alpha \in [0, \bar{\alpha}) \\ \text{Wrs}(\mathscr{B}_1) & , \forall \alpha \in [\bar{\alpha}, \pi] \end{cases} .$$

Proof. First it is easy to show that the worst-case evacuation time is induced either when $\mathscr{R}_1$ finds the exit while moving from cycle(0) to cycle(α), or while $\mathscr{R}_1, \mathscr{R}_2$ are exploring the circle together (after having met). By Lemma 2, the cost in the first case would be

$$\max_{0 \leq x \leq \alpha} \{1 + x + 2\delta_x\} = \begin{cases} 1 + \alpha + 2\delta_\alpha & , \text{if } \alpha \leq \alpha_0 \\ \text{Wrs}(\mathscr{B}_1) & , \text{otherwise} \end{cases}$$

where the values of the piecewise function above follow from Lemma 3. In the other case, the worst placement of exit is obtained using instances cycle$(\alpha + \epsilon)$ for arbitrary small values of $\epsilon > 0$ in which case the evacuation cost becomes $1 + 2\pi - \alpha$.

Overall, is is easy to see that $1 + \alpha_0 + 2\delta_{\alpha_0} \leq 1 + 2\pi - \alpha_0$ showing that the dominant evacuation cost when $\alpha \leq \bar{\alpha}$ is $1 + 2\pi - \alpha$. For $\alpha > \bar{\alpha}$ the evacuation cost becomes equal to w_1. $\quad\square$

In a similar fashion, we can easily analyze $\mathscr{A}_2$.

Lemma 5 (Worst-Case Analysis for $\mathscr{A}_2$). *For all* $\alpha \leq \pi - 2$, *we have* $\text{Wrs}(\mathscr{A}_2(\alpha)) = 1 + 2\pi - \alpha$.

Proof. We distinguish three cases as to where the exit is. If $x \in [0, \alpha)$, then the worst instance cycle(x) is when $x = \alpha - \epsilon$ for arbitrarily small $\epsilon > 0$, and the cost is $1 + \alpha + 2\delta_{\alpha/2}$. In the second case $x \in [\alpha, 2\alpha + \delta_{\alpha/2})$ and it is not difficult to see that the worst-case induced cost in this case is not more than that of the first case. Finally, in the third case $x \in [2\alpha + \delta_{\alpha/2}, 2\pi)$, and the two robots move together, so the total cost, in the worst-case, is $1 + 2\pi - \alpha$, when $x = 2\pi - \epsilon$ for arbitrarily small $\epsilon > 0$. It is not difficult to see that the dominant case is actually the third one, and in fact the two cases induce the same cost when $\pi = \alpha + \delta_{\alpha/2}$. By the definition of $\delta_{\alpha/2}$ we know that $\delta_{\alpha/2} = 2\sin\left(\frac{\alpha + \delta_{\alpha/2}}{2}\right) = 2\sin(\pi/2) = 2$. Hence, the costs become equal when $\alpha = \pi - 2$. $\quad\square$

Next, we analyze $\mathscr{A}_2'(\alpha, \beta)$, which requires more technical arguments. For this we will invoke $\mathscr{A}_2'$ only for special parameters, whose choice is motivated by the following observation pertaining to the performance of $\mathscr{A}_2$ (whose generalization is $\mathscr{A}_2'$). From the proof of Lemma 5, it follows that among all algorithms $\mathscr{A}_2(\alpha)$, where $\alpha \leq \gamma_0$ (see discussion before Definition 5), the one with minimum worst-case evacuation cost is $\mathscr{A}_2(\pi - 2)$, and the cost becomes $3 + \pi$. In fact, for all $w \in [3 + \pi, 1 + 2\pi]$ there are two different values of α for which $\text{Wrs}(\mathscr{A}_2(\alpha)) = w$, and we restrict $\alpha \in [0, \pi - 2]$ so that we obtain evacuation algorithms with minimum average-case cost. Moreover, $\alpha = \pi - 2$ is the only parameter for which $\text{Wrs}(\mathscr{A}_2(\alpha)) = 3 + \pi$ and as a byproduct, it is the algorithm in the family $\mathscr{A}_2$ that minimizes the worst-case.

By Lemma 5 we know that as $\beta \to 0$, the value of α that minimizes $\text{Wrs}(\mathscr{A}_2'(\alpha, \beta))$ approaches $\pi - 2$. That value of α is what made the evacuation cost of $\mathscr{A}_2(\alpha)$ attain the same value in two different (worst-case) exit placements. Motivated by this, and for values of $\beta > 0$ not too big, we still find the optimal choices of α that minimize the worst-case performance.

Lemma 6 (Worst-Case Analysis for $\mathscr{A}_2'$). *Let* $\beta_0 = 0.0438855$, *and set* $\alpha_\beta := \pi - \beta/2 - 2\cos(\beta/4)$. *Then for all* $\beta \in [0, \beta_0]$ *we have* $\text{Wrs}(\mathscr{A}_2'(\alpha_\beta, \beta)) = 1 + \pi - \beta/2 + 2\cos(\beta/4)$.

Proof. Let $w(\beta) = 1 + \pi - \beta/2 + 2\cos(\beta/4)$. First we show that $w(\beta)$ is the worst-case performance of $\mathscr{A}_2'(\alpha_\beta, \beta)$ for two specific placements of the exit.

We proceed by describing evacuation cost $\mathcal{C}(x)$ assuming two arbitrary α, β for two different instances cycle(x). Using Lemma 2, we see that

$$\lim_{\epsilon \to 0^+} \mathcal{C}(\alpha - \epsilon) = 1 + \lim_{\epsilon \to 0^+} \mathcal{S}(\alpha - \epsilon) + 2 \lim_{\epsilon \to 0^+} \mathcal{E}(\alpha - \epsilon) = 1 + \alpha + 2\delta_{\alpha/2}. \tag{3}$$

Since the total search time is $2\pi - \alpha - \beta$, we also see that

$$\lim_{\epsilon \to 0^+} \mathcal{C}(2\pi - \epsilon) = 1 + 2\pi - \alpha - \beta. \tag{4}$$

Now we claim that (3), (4) are equal when $\alpha = \alpha_\beta$. Indeed, equating (3), (4) gives

$$a + \delta_{\alpha/2} = \pi - \beta/2. \tag{5}$$

However, then, using (2), we see that

$$\delta_{\alpha/2} = 2\sin\left(\frac{\alpha + \delta_{\alpha/2}}{2}\right) = 2\sin\left(\frac{\pi - \beta/2}{2}\right) = 2\cos(\beta/4). \tag{6}$$

Substituting (6) into (5), we see that the value of α for which (3), (4) are equal satisfies $\alpha = \pi - \beta/2 - 2\cos(\beta/4)$, as promised. Substituting this special value of $\alpha = \alpha_\beta$ either in (3) or in (4) induces evacuation cost $w(\beta) = 1 + \pi - \beta/2 + 2\cos(\beta/4)$.

Next we show that as long as β is not too big, $w(\beta)$ is indeed the worst-case evacuation cost. We consider the following cases $x \in I_i, i = 1, \ldots, 4$ for possible instances cycle(x);

$$I_1 := [0, \alpha),$$
$$I_2 := [\alpha, 2\alpha + \beta + \delta_{(\alpha+\beta)/2}),$$
$$I_3 := [2\alpha + \beta + \delta_{(\alpha+\beta)/2}, 2\alpha + 2\beta + \delta_{(\alpha+\beta)/2} + \delta_{\beta/2}),$$
$$I_4 := [2\alpha + 2\beta + \delta_{(\alpha+\beta)/2} + \delta_{\beta/2}, 2\pi).$$

Clearly, (3), (4) demonstrate the worst-case evacuation costs for instances in I_1, I_4, respectively, and the cost in both cases, for $\alpha = \alpha_\beta$ is equal to $w(\beta)$.

If $x \in I_2$ then $\mathcal{C}(x) = 1 + \mathcal{S}(x) + 2\mathcal{E}(x)$. It is easy to see that both $\mathcal{S}(x), \mathcal{E}(x)$ are monotone in I_2, so the worst-case evacuation in this case is

$$\lim_{\epsilon \to 0^+} \mathcal{C}(2\alpha_\beta + \beta + \delta_{(\alpha_\beta + \beta)/2} - \epsilon) = 1 + \alpha_\beta + \beta + \delta_{(\alpha_\beta + \beta)/2} + 2\delta_{\beta/2}. \tag{7}$$

Denote $\delta_{\beta/2}$ satisfying (2) by δ_β'. Using (2) and the definition of α_β, we see that

$$\delta_{(\alpha_\beta + \beta)/2} = 2\sin\left(\frac{\alpha_\beta + \beta + \delta_{(\alpha_\beta + \beta)/2}}{2}\right) = 2\cos\left(\cos(\beta/4) - \beta/4 - \delta_{(\alpha_\beta + \beta)/2}\right).$$

For simplicity, we denote $\delta_{(\alpha_\beta + \beta)/2}$ that satisfies the equation above by δ_β''. Then, continuing from (7), the worst-case evacuation cost when $x \in I_2$ becomes $1 + \pi + \beta/2 - 2\cos(\beta/4) + \delta_\beta'' + 2\delta_\beta'$, an expression that depends exclusively on β. The latter cost is no more than $w(\beta)$ if and only if $4\cos(\beta/4) - \beta - \delta_\beta'' - 2\delta_\beta' \geq 0$, and numerically we verify that this is satisfied as long as $\beta \leq \beta_0$ (see also Figure 6).

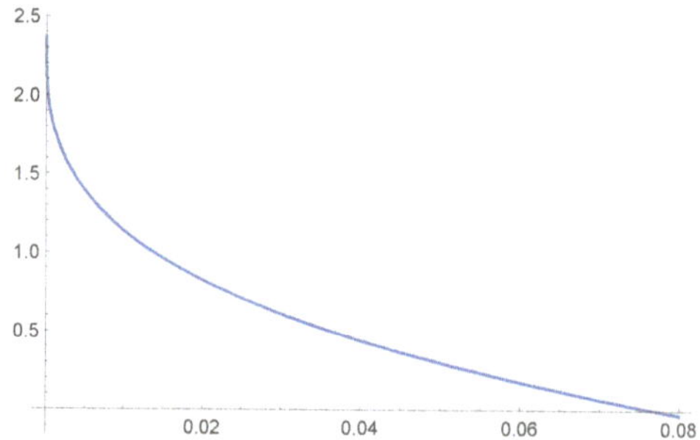

Figure 6. The behavior of expression $4\cos\left(\beta/4\right) - \beta - \delta_\beta'' - 2\delta_\beta'$, for $\beta = 0, \ldots, 0.8$.

Finally, it is easy to verify that $\delta_{\beta/2}$ and $|I_4|$ are increasing and decreasing, respectively, for $\beta \le \beta_0$, and that $\delta_{\beta_0/2} = 0.977997 \le 1.01099 = |I_4|$ (for $\beta = \beta_0$). As a result, the worst-case evacuation cost of case $x \in I_3$ cannot exceed that of case $x \in I_4$, and hence the lemma follows. $\square$

It is important to note that the worst-case performance of $\mathscr{A}_2'(\alpha_\beta, \beta)$ of Lemma 7 is decreasing in β. Indeed, $\frac{\partial}{\partial \beta}\mathrm{Wrs}\left(\mathscr{A}_2'(\alpha_\beta, \beta)\right) = \frac{1}{2} - \frac{1}{2}\sin\left(\beta/4\right) < 0$. For values of β close to 0, the derivative is nearly a constant. This also explains why in Figure 1, the performance of algorithm $\mathscr{A}_2'$ seems to have nearly invariant worst-case performance, which however is provably decreasing in β.

Lemma 7 (Worst-Case Analysis for $\mathscr{A}_2'$). *Let $\beta_0 = 0.0438855$, and set $\alpha_\beta := \pi - \beta/2 - 2\cos\left(\beta/4\right)$. Then for all $\beta \in [0, \beta_0]$ we have $\mathrm{Wrs}\left(\mathscr{A}_2'(\alpha_\beta, \beta)\right) = 1 + \pi - \beta/2 + 2\cos\left(\beta/4\right)$.*

6. Average-Case Performance Analysis and the Efficient Frontier

In this section, we perform average-case analysis for all algorithmic families $\mathscr{A}_1, \mathscr{A}_2, \mathscr{A}_2'$, with respect to their parameters. For the sake of exposition of our results, we set $w_1 = \mathrm{Wrs}\left(\mathscr{B}_1\right) \approx 5.73906$, $w_2 = \mathrm{Wrs}\left(\mathscr{B}_2\right) = 1 + 2\pi \approx 7.28319$ and for $\beta_0 \approx 0.04388$, as in Lemma 7, we set $w_0 := \mathrm{Wrs}\left(\mathscr{A}_2'(\alpha_{\beta_0}, \beta_0)\right) \approx 6.11953$. We also recall $\bar{\alpha} \approx 1.54419$ of Lemma 4. Finally, we set

$$v(\alpha) := 1 + 2\pi - \alpha$$
$$v_2(\beta) := 1 + \pi - \beta/2 + 2\cos\left(\beta/4\right)$$
$$u_1(\alpha) := 0.00889\alpha^3 - 0.16944\alpha^2 + 0.71518\alpha + 4.23089$$
$$u_2'(\beta) := 530.673\beta^3 - 78.5498\beta^2 + 7.36219\beta + 4.70493$$
$$u_2(\alpha) := 0.093056\alpha^2 + 0.346659\alpha + 4.1719$$

Combined with our findings of Section 5, the main result of the current section is the following.

Theorem 6. *For every $w \in [w_1, w_2]$ there is algorithm $\mathscr{A} \in \{\mathscr{A}_1, \mathscr{A}_2', \mathscr{A}_2\}$ and unique parameter(s) p such that $\mathrm{Wrs}\left(\mathscr{A}(p)\right) = w$. In particular,*

- *for all $\alpha \in [1, \bar{\alpha}]$, $\mathscr{A}_1(\alpha)$ is $(u_1(\alpha), v(\alpha))$-efficient, and $v([1, \bar{\alpha}]) = [w_1, 2\pi]$,*
- *for all $\beta \in [0, \beta_0]$, $\mathscr{A}_2'(\alpha_\beta, \beta)$ is $(u_2'(\beta), v_2(\beta))$-efficient, and $v_2([0, \beta_0]) = [w_0, 3 + \pi]$,*
- *for all $\alpha \in [0, \pi - 2]$, $\mathscr{A}_2(\alpha)$ is $(u_2(\alpha), v(\alpha))$-efficient, and $v([0, \pi - 2]) = [3 + \pi, w_2]$.*

Proof. The claims for the worst-case performances of $\mathscr{A}_1, \mathscr{A}_2', \mathscr{A}_2$ follow directly from Lemmata 4, 7 and 5, respectively. Next we argue that as the parameters vary in their specified range, we obtain the entire spectrum of $w \in [w_1, w_2]$, and this for unique values of the parameters. For this, we will rely on that for all evacuation algorithm families, the worst-case cost is monotone in the parameters.

First, we argue about $\mathscr{A}_1$. We observe that by the definition of $\bar{\alpha}$, $\mathrm{Wrs}\left(\mathscr{A}_1(\bar{\alpha})\right) = w_1$, and $\mathrm{Wrs}\left(\mathscr{A}_1(1)\right) = 1 + 2\pi - 1 = 2\pi$. Together with the fact that $v(\alpha)$ is strictly decreasing, we see that $\mathrm{Wrs}\left(\mathscr{A}_1(\alpha)\right)$ is 1-1 and onto to $[w_1, 2\pi]$ as α ranges in $[1, \bar{\alpha}]$.

Second, we study $\mathscr{A}'_2$ whose worst-case cost $v_2(\beta)$ is strictly decreasing in β. Moreover, by definition of β_0, we have $\mathrm{Wrs}\left(\mathscr{A}_2(\alpha_{\beta_0}, \beta_0)\right) = w_0$. Then we note that for $\beta = 0$, $\mathscr{A}_2(\alpha_\beta, \beta)$ coincides with $\mathscr{A}_2(\pi - 2)$, and in particular the induced worst-case cost becomes $3 + \pi$. Therefore $\mathrm{Wrs}\left(\mathscr{A}'_2(\alpha_\beta, \beta)\right)$ is 1-1 and onto to $[w_0, 3 + \pi]$ as β ranges in $[0, \beta_0]$.

Third, we study $\mathscr{A}_2$, for which we know that $\mathrm{Wrs}\left(\mathscr{A}_2(\pi - 2)\right) = 3 + \pi$. Again, the worst-case cost is monotone in α and $\mathscr{A}_2(0)$ coincides with benchmark algorithm $\mathscr{B}_2$, which is $\mathrm{Wrs}\left(\mathscr{A}_2(0)\right) = w_2$. Hence, $\mathrm{Wrs}\left(\mathscr{A}_2(\alpha)\right)$ is 1-1 and onto to $[3 + \pi, w_2]$ as α ranges in $[0, \pi - 2]$.

Finally, we argue that

$$\mathrm{Avg}\left(\mathscr{A}_1(\alpha)\right) \leq u_1(\alpha), \forall \alpha \in [1, \bar{\alpha}]$$

$$\mathrm{Avg}\left(\mathscr{A}'_2(\alpha_\beta, \beta)\right) \leq u'_2(\beta), \forall \beta \in [0, \beta_0]$$

$$\mathrm{Avg}\left(\mathscr{A}_2(\alpha)\right) \leq u_2(\alpha), \forall \alpha \in [0, \pi - 2]$$

For this, we numerically compute $\mathrm{Avg}\left(\mathscr{A}_1(\alpha)\right), \mathrm{Avg}\left(\mathscr{A}'_2(\alpha_\beta, \beta)\right), \mathrm{Avg}\left(\mathscr{A}_2(\alpha)\right)$ for various values of parameters α, β, and we heuristically choose u_1, u'_2, u_2 so as to upper bound the average-case performance of $\mathscr{A}_1, \mathscr{A}'_2, \mathscr{A}_2$, effectively verifying our claim numerically. For each evacuation algorithm, we use Corollary 1, which together with the analytic description of our evacuation algorithms (see Definitions 4, 6, and 5) allow us to compute their average-case performance using computer-assisted calculations. Our numerical calculations are depicted in Figure 7.

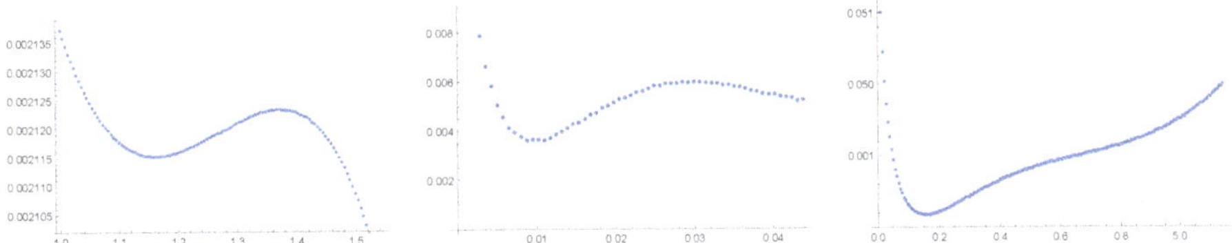

Figure 7. On the right $u_1(\alpha) - \mathrm{Avg}\left(\mathscr{A}_1(\alpha)\right)$, for $\alpha' \leq \alpha \leq \bar{\alpha}$. In the middle, $u'_2(\beta) - \mathrm{Avg}\left(\mathscr{A}'_2(\alpha_\beta, \beta)\right)$, for $0 \leq \beta \leq \beta_0$. On the right $u_2(\alpha) - \mathrm{Avg}\left(\mathscr{A}_2(\alpha)\right)$, for $0 \leq \alpha \leq \pi - 2$.

$\square$

Finally, we aim to formally quantify the efficient frontier of our algorithms as depicted in Figure 1 (see Section 2.1). The parametric curves described in Theorem 6 provide, strictly speaking, an upper bound for the parametric curve of Figure 1. Next, we compute $g : \mathbb{R} \mapsto \mathbb{R}$, so that the parametric curves of Theorem 6 are written in the form $\{(g(w), w)\}_{w \in [w_1, w_2]}$. That would also imply that there is a solution to $_2\mathrm{EVAC}^w_{F2F}$ of cost at most $g(w)$.

In that direction, we study each evacuation algorithm family $\mathscr{A}(p)$ with worst-case performance, say, $v(p)$, and average-case upper bound, say, $u(p)$. For each $w \in [w_1, w_2]$ in the range of $\mathscr{A}(p)$, we set $p = v^{-1}(w)$ so that the average-case performance achieved becomes $u(v^{-1}(w))$.

Recall that $\mathrm{Wrs}\left(\mathscr{A}_i(\alpha)\right) = v(\alpha)$, so that $v^{-1}(w) = 1 + 2\pi - w$, and hence for algorithms $\mathscr{A}_i$ we can easily compute $u_i(v^{-1}(w)), i = 1, 2$. For $\mathscr{A}'_2$ we recall that $\mathrm{Avg}\left(\mathscr{A}'_2(\alpha_\beta, \beta)\right)$ is decreasing in β. Since v_2^{-1} does not admit a closed form, we need to observe that $2.999 + \pi - \beta/2 \leq v_2(\beta) \leq 3 + \pi - \beta/2$ for all $\beta \in [0, \beta_0]$ so that an upper bound for $\mathrm{Avg}\left(\mathscr{A}'_2(\alpha_\beta, \beta)\right)$ admitting worst-case performance w can be computed by $u'_2(12.2812 - 2w)$.

Now for each $w \in [w_1, w_2]$ we need to specify which of the evacuation algorithms we will invoke. Please note that in Theorem 6 we chose the range of α in $\mathscr{A}_1$ to start from 1 so that as to guarantee that $\mathrm{Wrs}\left(\mathscr{A}_1(1)\right) \geq w_0$. We note that $u'_2(12.2812 - 2w) = u_1(1 + 2\pi - w)$ for $w' \approx 6.12851$, so algorithm $\mathscr{A}_1$ should be invoked for $w \in [w_1, w']$ (and w' is obtained for $\alpha' := 1 + 2\pi - w' \approx 1.15468$), then $\mathscr{A}'_2$

for $w \in [w', 3 + \pi]$ (and w' is obtained for β' so that $v_2(\beta') = w'$, where $\beta' \approx 0.0241653$), and $\mathscr{A}_2$ for $w \in [3 + \pi, w_2]$. We conclude with the next Theorem (for convenience, the values of all constants are summarized at the end of Section 2.3).

Theorem 7. *For every $w \in [w_1, w_2]$, the optimal solution to $_2\text{EVAC}^w_{F2F}$ is at most $g(w)$, where*

$$g(w) = \begin{cases} -0.00889w^3 + 0.0248026w^2 + 0.338241w + 3.88629 & , w \in [w_1, w'] \ (\mathscr{A}_1(\alpha), \ \alpha \in [\alpha', \bar{\alpha}]) \\ -4245.38w^3 + 77893.3w^2 - 476397.w + 971235 & , w \in [w', 3 + \pi] \ (\mathscr{A}_2(\alpha_\beta, \beta), \ \beta \in [0, \beta']) \\ 0.093056w^2 - 1.70215w + 11.6328 & , w \in [3 + \pi, w_2] \ (\mathscr{A}_2(\alpha), \ \alpha \in [0, \pi - 2]) \end{cases}$$

7. Conclusions and Open Problems

We offered a new perspective to the well-studied problem of evacuating two robots from the disk, in the face-to-face model, first introduced in [1]. A series of results pertaining especially to the same domain have focused exclusively on deterministic algorithms and their worst-case (competitive) analyses. Our work can be understood as a first attempt to study the same problem in the realm of randomized algorithms. More specifically, in light of known positive results for the problem, we asked the question of minimizing the average-case performance of an evacuation algorithm, condition on that the worst-case performance remains bounded. We allowed that latter bound to range between the best known guarantee of a simple yet efficient evacuation algorithm (that was introduced in [1]) and that of a naive algorithm that simulates the optimal solution of one searcher. Our main contribution is the introduction of three new evacuation algorithms that perform well for different values of the worst-case bound, inducing a continuous bound of the efficient (pareto) frontier for the underlying multi-objective optimization problem (that of minimizing both the worst and the average-case performance). We also motivated our results by verifying, somehow surprisingly, that among the family of algorithms introduced in [27], which gave the second subsequent improvement of the worst-case performance, none of them could improve the average-case performance of the simple algorithm presented in [1]. In that sense our new algorithms outperform the best algorithms known for the problem, but in the new multi-objective setting. Our findings give rise to several open questions and directions that aim to better understand $_2\text{EVAC}^w_{F2F}$ as well as multi-objective optimization search-type problems.

Open questions specific to our problem $_2\text{EVAC}^w_{F2F}$ are:

- Prove lower bounds for $_2\text{EVAC}^w_{F2F}$, for any w. Is any of our algorithms, for any w optimal?
- For the value $w = \text{Wrs}\,(\mathscr{B}_1)$, we designed algorithm $\mathscr{A}_1(\alpha)$ for $_2\text{EVAC}^w_{F2F}$ which for a proper value of a has worst-case performance exactly w, while its average-case performance is strictly less than $\text{Avg}\,(\mathscr{B}_1)$. Is it feasible to attain worst-case performance strictly less than w, while having average-case performance at most $\text{Avg}\,(\mathscr{B}_1)$?
- The bound to the efficient (pareto) frontier we derived for problem $_2\text{EVAC}^w_{F2F}$ is indeed continuous, with respect to parameter w, but not differentiable. Is the optimal pareto frontier smooth, or is there any other family of algorithms that improves upon our results and gives a smooth transition between families of evacuation algorithms?
- The algorithmic families we derived for $_2\text{EVAC}^w_{F2F}$ exhibit the following property. $\mathscr{A}_2$ is a natural extension to $\mathscr{B}_2$. Similarly, $\mathscr{A}_2'$ is a natural extension to $\mathscr{A}_2$. Finally, $\mathscr{A}_1$ is a natural extension to $\mathscr{B}_1$. However, $\mathscr{A}_1$ and $\mathscr{A}_2'$ have different behavior (there are no values of their parameters that induce the same evacuation protocol), even though for a proper choice of their parameters, they induce algorithms with the same worst-case and average-case performance.
- Observe that the average-case performance of $\mathscr{B}_2$ is $1 + \pi$. All our evacuation algorithms induce average cost at least $1 + \pi$. We conjecture that even in the wireless model, as well as for any number of robots, $1 + \pi$ is tight lower bound for the average performance of evacuation algorithm.

We conclude with some future directions that are inspired by our work and pertain to more general problems:

- Our algorithms can also be interpreted as randomized algorithms that have access to infinitely many bits (or enough many bits, in order to simulate a uniformly random deployment point on the circle). What if the algorithm has access only to a limited number of random bits?
- To the best of our knowledge, the current paper is the first attempt to study multi-objective optimization search-type problems. It was followed by [48,49] who considered time energy trade-offs for a search problems on the line. This line of research admits many future directions based on any combination of multiple objectives, e.g., worst-case, average-case and competitive cost, time, energy and any other efficiency measure, or even trade-offs involving number of faults or even complexity resources, e.g., memory, communication or randomness.

Author Contributions: Investigation, H.C. and P.S.; Supervision, K.G. All authors have read and agreed to the published version of the manuscript.

Funding: The second author and this research was supported in part by NSERC Discovery Grant.

Conflicts of Interest: The authors declare no conflict of interest.

References

1. Czyzowicz, J.; Gasieniec, L.; Gorry, T.; Kranakis, E.; Martin, R.; Pajak, D. Evacuating Robots via Unknown Exit in a Disk. In Proceedings of the DISC, Austin, TX, USA, 12–15 October 2014; Springer: Berlin/Heidelberg, Germany, 2014; pp. 122–136.
2. Beck, A. On the linear search problem. *Isr. J. Math.* **1964**, *2*, 221–228. [CrossRef]
3. Bellman, R. An optimal search. *SIAM Rev.* **1963**, *5*, 274–274. [CrossRef]
4. Dobbie, J. A survey of search theory. *Oper. Res.* **1968**, *16*, 525–537. [CrossRef]
5. Benkoski, S.; Monticino, M.; Weisinger, J. A survey of the search theory literature. *Nav. Res. Logist. (NRL)* **1991**, *38*, 469–494. [CrossRef]
6. Stone, L. *Theory of Optimal Search*; Academic Press: New York, NY, USA, 1975.
7. Ahlswede, R.; Wegener, I. *Search Problems*; Wiley-Interscience: New York, NY, USA, 1987.
8. Alpern, S.; Gal, S. *The Theory of Search Games and Rendezvous*; Kluwer Academic Publishers: Dordrecht, The Netherlands, 2002; Volume 55.
9. Alpern, S.; Fokkink, R.; Gasieniec, L.; Lindelauf, R.; Subrahmanian, V. (Eds.) Ten Open Problems in Rendezvous Search. In *Search Theory: A Game Theoretic Perspective*; Springer: New York, NY, USA, 2013; pp. 223–230.
10. Albers, S.; Henzinger, M.R. Exploring unknown environments. *SIAM J. Comput.* **2000**, *29*, 1164–1188. [CrossRef]
11. Albers, S.; Kursawe, K.; Schuierer, S. Exploring unknown environments with obstacles. *Algorithmica* **2002**, *32*, 123–143. [CrossRef]
12. Deng, X.; Kameda, T.; Papadimitriou, C. How to learn an unknown environment. In Proceedings of the 32nd Annual Symposium of Foundations of Computer Science, San Juan, Puerto Rico, 1–4 October 1991; pp. 298–303.
13. Hoffmann, F.; Icking, C.; Klein, R.; Kriegel, K. The polygon exploration problem. *SIAM J. Comput.* **2001**, *31*, 577–600. [CrossRef]
14. Burgard, W.; Moors, M.; Stachniss, C.; Schneider, F.E. Coordinated multi-robot exploration. *IEEE Trans. Robot.* **2005**, *21*, 376–386. [CrossRef]
15. Thrun, S. A probabilistic on-line mapping algorithm for teams of mobile robots. *Int. J. Robot. Res.* **2001**, *20*, 335–363. [CrossRef]
16. Yamauchi, B. Frontier-based exploration using multiple robots. In Proceedings of the Second International Conference on Autonomous Agents, Minneapolis, MI, USA, 9–13 May 1998; ACM: New York, NY, USA, 1998; pp. 47–53.
17. Kleinberg, J. On-line search in a simple polygon. In Proceedings of the SODA, Arlington, VA, USA, 23–25 January 1994; p. 8.

18. Mitchell, J.S. Geometric shortest paths and network optimization. *Handb. Comput. Geom.* **2000**, *334*, 633–702.

19. Papadimitriou, C.H.; Yannakakis, M. Shortest paths without a map. In *ICALP*; Springer: Berlin/Heidelberg, Germany, 1989; pp. 610–620.

20. Chung, T.H.; Hollinger, G.A.; Isler, V. Search and pursuit-evasion in mobile robotics. *Auton. Robot.* **2011**, *31*, 299–316.

21. Fomin, F.V.; Thilikos, D.M. An annotated bibliography on guaranteed graph searching. *Theor. Comput. Sci.* **2008**, *399*, 236–245.

22. Lidbetter, T. Hide-and-Seek and other Search Games. Ph.D. Thesis, The London School of Ecoomics and Political Science (LSE), London, UK, 2013.

23. Nahin, P. *Chases and Escapes: The Mathematics of Pursuit and Evasion*; Princeton University Press: Princeton, NJ, USA, 2012.

24. Baumann, N.; Skutella, M. Earliest arrival flows with multiple sources. *Math. Oper. Res.* **2009**, *34*, 499–512.

25. Fekete, S.; Gray, C.; Kröller, A. Evacuation of rectilinear polygons. In *Combinatorial Optimization and Applications*; Springer: Berlin/Heidelberg, Germany, 2010; pp. 21–30.

26. Czyzowicz, J.; Georgiou, K.; Kranakis, E.; Narayanan, L.; Opatrny, J.; Vogtenhuber, B. Evacuating Robots from a Disc Using Face to Face Communication. In Proceedings of the CIAC 2015, Paris, France, 20–22 May 2015; Springer: Berlin/Heidelberg, Germany, 2015.

27. Brandt, S.; Laufenberg, F.; Lv, Y.; Stolz, D.; Wattenhofer, R. Collaboration without Communication: Evacuating Two Robots from a Disk. In Proceedings of the Algorithms and Complexity—10th International Conference, CIAC, Athens, Greece, 24–26 May 2017; pp. 104–115.

28. Disser, Y.; Schmitt, S. Evacuating two robots from a disk: A second cut. In *International Colloquium on Structural Information and Communication Complexity*; Lecture Notes in Computer Science; Censor-Hillel, K., Flammini, M., Eds.; Springer: Berlin/Heidelberg, Germany, 2019; Volume 11639, pp. 200–214.

29. Czyzowicz, J.; Dobrev, S.; Georgiou, K.; Kranakis, E.; MacQuarrie, F. Evacuating two robots from multiple unknown exits in a circle. *Theor. Comput. Sci.* **2018**, *709*, 20–30.

30. Pattanayak, D.; Ramesh, H.; Mandal, P.S.; Schmid, S. Evacuating Two Robots from Two Unknown Exits on the Perimeter of a Disk with Wireless Communication. In Proceedings of the 19th International Conference on Distributed Computing and Networking, ICDCN 2018, Varanasi, India, 4–7 January 2018; Bellavista, P., Garg, V.K., Eds.; ACM: New York, NY, USA, 2018; pp. 20:1–20:4.

31. Chuangpishit, H.; Mehrabi, S.; Narayanan, L.; Opatrny, J. Evacuating equilateral triangles and squares in the face-to-face model. *Comput. Geom.* **2020**, *89*, 101624.

32. Czyzowicz, J.; Kranakis, E.; Krizanc, D.; Narayanan, L.; Opatrny, J.; Shende, S. Wireless Autonomous Robot Evacuation from Equilateral Triangles and Squares. In Proceedings of Ad-hoc, Mobile, and Wireless Networks, ADHOC-NOW, Athens, Greece, 29 June–1 July 2015; pp. 181–194.

33. Brandt, S.; Foerster, K.T.; Richner, B.; Wattenhofer, R. Wireless evacuation on m rays with k searchers. *Theor. Comput. Sci.* **2020**, *811*, 56–69.

34. Angelopoulos, S.; Dürr, C.; Lidbetter, T. The expanding search ratio of a graph. *Discret. Appl. Math.* **2019**, *260*, 51–65.

35. Borowiecki, P.; Das, D.; Dereniowski, D.; Kuszner, L. Distributed Evacuation in Graphs with Multiple Exits. In *Structural Information and Communication Complexity, Proceedings of the 23rd International Colloquium, SIROCCO 2016, Helsinki, Finland, 19–21 July 2016*; Revised Selected Papers, Lecture Notes in Computer Science; Suomela, J., Ed.; Springer: Berlin/Heidelberg, Germany, 2016; Volume 9988, pp. 228–241.

36. Chrobak, M.; Gasieniec, L.T.G.; Martin, R. Group Search on the Line. In *SOFSEM 2015*; Springer: Berlin/Heidelberg, Germany, 2015.

37. Georgiou, K.; Lucier, J. Weighted Group Search on a Line. In Proceedings of the 16th International Symposium on Algorithms and Experiments for Wireless Sensor Networks, ALGOSENSORS, Pisa, Italy, 7–11 September 2020.

38. Baeza Yates, R.; Culberson, J.; Rawlins, G. Searching in the plane. *Inf. Comput.* **1993**, *106*, 234–252.

39. Czyzowicz, J.; Georgiou, K.; Godon, M.; Kranakis, E.; Krizanc, D.; Rytter, W.; Włodarczyk, M. Evacuation from a disc in the presence of a faulty robot. In *International Colloquium on Structural Information and Communication Complexity*; Springer: Berlin/Heidelberg, Germany, 2017; pp. 158–173.

40. Georgiou, K.; Kranakis, E.; Leonardos, N.; Pagourtzis, A.; Papaioannou, I. Optimal Cycle Search Despite the Presence of Faulty Robots. In *Algorithms for Sensor Systems, Proceedings of the 15th International Symposium on Algorithms and Experiments for Wireless Sensor Networks, ALGOSENSORS 2019, Munich, Germany, 12–13 September 2019*; Revised Selected Papers, Lecture Notes in Computer Science; Dressler, F., Scheideler, C., Eds.; Springer: Berlin/Heidelberg, Germany, 2019; Volume 11931, pp. 192–205.

41. Pattanayak, D.; Ramesh, H.; Mandal, P.S. Chauffeuring a Crashed Robot from a Disk. In *Algorithms for Sensor Systems, Proceedings of the 15th International Symposium on Algorithms and Experiments for Wireless Sensor Networks, ALGOSENSORS 2019, Munich, Germany, 12–13 September 2019*; Revised Selected Papers, Lecture Notes in Computer Science; Dressler, F., Scheideler, C., Eds.; Springer: Berlin/Heidelberg, Germany, 2019; Volume 11931, pp. 177–191.

42. Bonato, A.; Georgiou, K.; MacRury, C.; Pralat, P. Probabilistically Faulty Searching on a Half-Line. In Proceedings of the 14th Latin American Theoretical Informatics Sumposium, University of Sao Paulo, Sao Paulo, Brazil, 25–29 May 2020.

43. Georgiou, K.; Kranakis, E.; Steau, A. Searching with Advice: Robot Fence-Jumping. *J. Inf. Process.* **2017**, *25*, 559–571.

44. Czyzowicz, J.; Georgiou, K.; Killick, R.; Kranakis, E.; Krizanc, D.; Narayanan, L.; Opatrny, J.; Shende, S. Priority Evacuation from a Disk: The case of $n \geq 4$. *Theor. Comput. Sci.* **2020**, accepted.

45. Czyzowicz, J.; Georgiou, K.; Killick, R.; Kranakis, E.; Krizanc, D.; Narayanan, L.; Opatrny, J.; Shende, S.M. Priority evacuation from a disk: The case of n = 1, 2, 3. *Theor. Comput. Sci.* **2020**, *806*, 595–616. [CrossRef]

46. Georgiou, K.; Karakostas, G.; Kranakis, E. Search-and-Fetch with One Robot on a Disk—(Track: Wireless and Geometry). In *Algorithms for Sensor Systems, Proceedings of the 12th International Symposium on Algorithms and Experiments for Wireless Sensor Networks, ALGOSENSORS 2016, Aarhus, Denmark, 25–26 August 2016*; Revised Selected Papers; Springer: Berlin/Heidelberg, Germany, 2016; pp. 80–94.

47. Georgiou, K.; Karakostas, G.; Kranakis, E. Search-and-Fetch with 2 Robots on a Disk: Wireless and Face-to-Face Communication Models. *Discret. Math. Theor. Comput. Sci.* **2019**, *21*. [CrossRef]

48. Czyzowicz, J.; Georgiou, K.; Killick, R.; Kranakis, E.; Krizanc, D.; Lafond, M.; Narayanan, L.; Opatrny, J.; Shende, S. Energy Consumption of Group Search on a Line. In *Leibniz International Proceedings in Informatics (LIPIcs), Proceedings of the 46th International Colloquium on Automata, Languages, and Programming (ICALP 2019), Patras, Greece, 8–12 July 2019*; Baier, C., Chatzigiannakis, I., Flocchini, P., Leonardi, S., Eds.; Schloss Dagstuhl–Leibniz-Zentrum fuer Informatik: Dagstuhl, Germany, 2019; Volume 132, pp. 137:1–137:15. [CrossRef]

49. Kranakis, E.J.; Krizanc, D.K.; Georgiou, M.L.; Killick, R.; Narayanan, L.; Opatrny, J.; Shende, S. Time-Energy Tradeoffs for Evacuation by Two Robots in the Wireless Model. In *Structural Information and Communication Complexity, Proceedings of the 26th International Colloquium, SIROCCO 2019, L'Aquila, Italy, 1–4 July 2019*; Lecture Notes in Computer Science; Censor-Hillel, K., Flammini, M., Eds.; Springer: Berlin/Heidelberg, Germany, 2019; Volume 11639, pp. 185–199.

50. Lamprou, I.; Martin, R.; Schewe, S. Fast two-robot disk evacuation with wireless communication. In Proceedings of the DISC, Paris, France, 27–29 September 2016; pp. 1–15.

51. Czyzowicz, J.; Kranakis, E.; Krizanc, D.; Narayanan, L.; Opatrny, J.; Shende, S.M. Linear Search with Terrain-Dependent Speeds. In *Algorithms and Complexity, Proceedings of the 10th International Conference, CIAC 2017, Athens, Greece, 24–26 May 2017*; Lecture Notes in Computer Science; Fotakis, D., Pagourtzis, A., Paschos, V.T., Eds.; Springer: Berlin/Heidelberg, Germany, 2017; Volume 10236, pp. 430–441.

52. Czyzowicz, J.; Georgiou, K.; Kranakis, E. Group Search and Evacuation. In *Distributed Computing by Mobile Entities; Current Research in Moving and Computing*; Flocchini, P., Prencipe, G., Santoro, N., Eds.; Springer: Berlin/Heidelberg, Germany, 2019; Chapter 14, pp. 335–370.

53. López-Ortiz, A.; Sweet, G. Parallel searching on a lattice. In Proceedings of the CCCG, 13–15 August 2001; pp. 125–128.

54. Emek, Y.; Langner, T.; Uitto, J.; Wattenhofer, R. Solving the ANTS Problem with Asynchronous Finite State Machines. In *Automata, Languages, and Programming, Proceedings of the 41st International Colloquium, ICALP 2014, Copenhagen, Denmark, 8–11 July 2014*; Lecture Notes in Computer Science, Part II; Esparza, J., Fraigniaud, P., Husfeldt, T., Koutsoupias, E., Eds.; Springer: Berlin/Heidelberg, Germany, 2014; Volume 8573, pp. 471–482.

55. Lenzen, C.; Lynch, N.; Newport, C.; Radeva, T. Trade-offs Between Selection Complexity and Performance when Searching the Plane Without Communication. In *ACM Symposium on Principles of Distributed Computing, Proceedings of the PODC '14, Paris, France, 15–18 July 2014*; HalldÃ³rsson, M.M., Dolev, S., Eds.; ACM: New York, NY, USA, 2014; pp. 252–261.

56. Acharjee, S.; Georgiou, K.; Kundu, S.; Srinivasan, A. Lower Bounds for Shoreline Searching With 2 or More Robots. In Proceedings of the 23rd International Conference on Principles of Distributed Systems (OPODIS'19), Neuchâtel, Switzerland, 17–19 December 2019; Felber, P., Friedman, R., Gilbert, S., Miller, A., Eds.; Schloss Dagstuhl—Leibniz-Zentrum fur Informatik: Dagstuhl, Germany, 2019; Volume 153, pp. 26:1–26:11.

57. Dobrev, S.; Kralovic, R.; Pardubska, D. Improved Lower Bounds for Shoreline Search. In *Structural Information and Communication Complexity, Proceedings of the 27th International Colloquium, SIROCCO 2020, Paderborn, Germany, 29 June–1 July 2020*; Lecture Notes in Computer Science; Springer: Berlin/Heidelberg, Germany, 2020.

58. Miettinen, K. Nonlinear multiobjective optimization. In *International Series in Operations Research and Management Science*; Kluwer: Dordrecht, The Netherlands, 1998; Volume 12, pp. 1–298.

Publisher's Note: MDPI stays neutral with regard to jurisdictional claims in published maps and institutional affiliations.

MDPI

St. Alban-Anlage 66

4052 Basel

Switzerland

Tel. +41 61 683 77 34

Fax +41 61 302 89 18

www.mdpi.com

Information Editorial Office

E-mail: information@mdpi.com

www.mdpi.com/journal/information